AF704232

region4
Educated Solutions

Gateways to Science

STAAR Edition

Grade 5

Student Edition

Region 4 Education Service Center supports student achievement by providing educational products and services that focus on excellence, service, and children.

Published by
Region 4 Education Service Center
7145 West Tidwell Road
Houston, Texas 77092-2096
www.esc4.net

© 2006, 2011 by Region 4 Education Service Center. All rights reserved. Except as permitted under the United States Copyright Act of 1976, no part of this publication may be reproduced or distributed in any form or by any means, or stored in a database or retrieval system, without the prior written permission of the publisher.

First edition 2006

Second edition 2011

18 17 16 15 14 13 12 11 2 3 4 5

ISBN-13: 978-1-934950-90-6

Printed in the United States of America

Acknowledgments

Region 4 Education Service Center would like to acknowledge the talent and expertise of those who contributed to the development of this book. Their dedication to our core values of excellence, service, and children made possible the creation of this resource to assist educators in providing quality, effective instruction for all students.

Writing Team
Janis Massey Macomber
Hanna K. Roach
Anne Douglas
Anna McClane

Design Team
Shelly D. Smith
Delia Ornelas
Lenworth Johnson
Billy Sanders
David Martinez

Editing Team
Jessica Hilbert
Brenda Gunter
Lisa Roys
Becky Sanders

TABLE of Contents

© REGION 4 EDUCATION SERVICE CENTER. ALL RIGHTS RESERVED.

TABLE of Contents

Unit 1

Thinking like a Scientist

UNIT 1: Thinking like a Scientist

A Word to the Student

Gateways to Science for Grade 5 is a textbook written for fifth-grade scientists. Are you a scientist? Yes, you are. A scientist reads, studies, and does science. This year you will experience hands-on activities, interesting reading passages, and group investigations. Get ready to think, plan, question, explore, and observe the world of science.

In *Gateways to Science*, you will notice that the lessons are planned in a special way. Each lesson has five parts. These five parts are called the 5 Es.

Let's Engage!

The first E begins a new lesson. In this activity, you link something you already know to something you are about to learn.

Let's Explore!

The second E actively explores what you are learning. Most likely, you work in a group. The group investigates a science topic and discusses its findings.

Let's Explain!

The third E is a time of discussion and sharing with your classmates and teacher. During this time, you talk about questions arising from your findings. You may also write in your science notebook about your new understanding.

Let's Elaborate!

In thc fourth E, you takc what you know and connect it to a new situation. You may read to learn more about the topic, create a product, or complete another group investigation.

Let's Evaluate!

The fifth E gives you the chance to show what you have learned and are able to do. This is a time when you and your teacher can check your progress.

You have a lot to look forward to.

Working in a group may be a new experience for you. To be successful, students in a group must work as a team. Teamwork includes respecting each other and each other's ideas. What does that mean? Listen when someone speaks. Disagree, if necessary, in a polite manner. Work together to solve problems.

Using a science notebook may also be a new experience for you. In *Gateways to Science*, writing to learn is important. You will record your ideas and observations in your science notebook. Your notebook may also include drawings, tables, and charts.

To be a good scientist, you must be willing to ask questions. Be brave! Be courageous! Ask questions if you do not understand something. Will you always get your answer right away? Maybe. Maybe not. Discovering answers to questions is part of doing science.

Finally, *Gateways to Science* is your guide to investigating the worlds of science. Get ready for the experience.

© REGION 4 EDUCATION SERVICE CENTER. ALL RIGHTS RESERVED.

UNIT 1: Thinking like a Scientist

Lesson 1: Safety

Let's Engage!

1. 

2. 

3. 

4. 

5. 

What does each sign mean? What do all the signs have in common?

Let's Explore!

Brainstorm answers to the question with your group:

What safety rules or guidelines do you follow?

Let's Explain!

1. Read the statements from the Science Laboratory Safety Contract.

Science Laboratory Safety Contract

- I will act responsibly at all times in the science laboratory and classroom.
- I will follow all instructions given by the teacher about lab procedures.
- I will keep my work area clean and organized.
- I will wear safety goggles and protective clothing, when necessary, at all times.
- I will know what to do in case of a fire.
- I will immediately notify the teacher of any emergency.
- I will tie back long hair, remove dangling jewelry, and secure loose clothing.
- I will not take materials or equipment out of the lab without direct permission from the teacher.
- I will never eat or drink in the laboratory unless instructed to do so by the teacher.
- I will handle living organisms or preserved specimens only when authorized by the teacher.
- I will not enter the storage area unless directed by the teacher.
- I will follow classroom guidelines for the conservation, disposal, and recycling of materials.

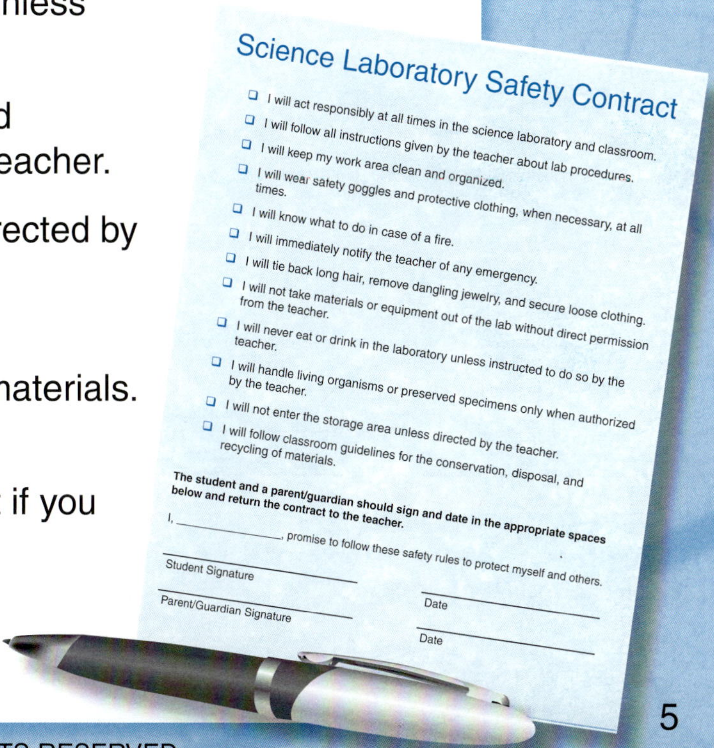

2. Ask your teacher to explain a statement if you do not understand it.

© REGION 4 EDUCATION SERVICE CENTER. ALL RIGHTS RESERVED.

3. Sign the copy of the Science Laboratory Safety Contract that your teacher provides for you.

4. Take the copy home and have a parent or guardian read and sign the contract.

5. Return the signed contract to your teacher.

Let's Elaborate!

As a group—

- Draw a card from the container.
- Read the card.
- Discuss answers to the questions.
- Be ready to share your answers.

Let's Evaluate!

Use your knowledge of safety to complete the assessment.

Let's Engage!

A Bug-Free Picnic

Scientists in New Zealand have made paper cartons that keep bugs away. The cartons are coated with oil made from grasses. The oil is called citronella.

The scientists tested the cartons. They coated five paper cartons with different types of bug repellent. One was coated with citronella. Some of the paper cartons had nothing on them at all.

The scientists let loose hundreds of beetles near the cartons. The beetles stayed away from the carton that was coated with citronella oil. The scientists repeated the test many times. The results of the test were reliable.

This may be the first step to keeping bugs out of cartons of food. Scientists must now be sure that paper cartons coated with citronella are safe to store food for humans.

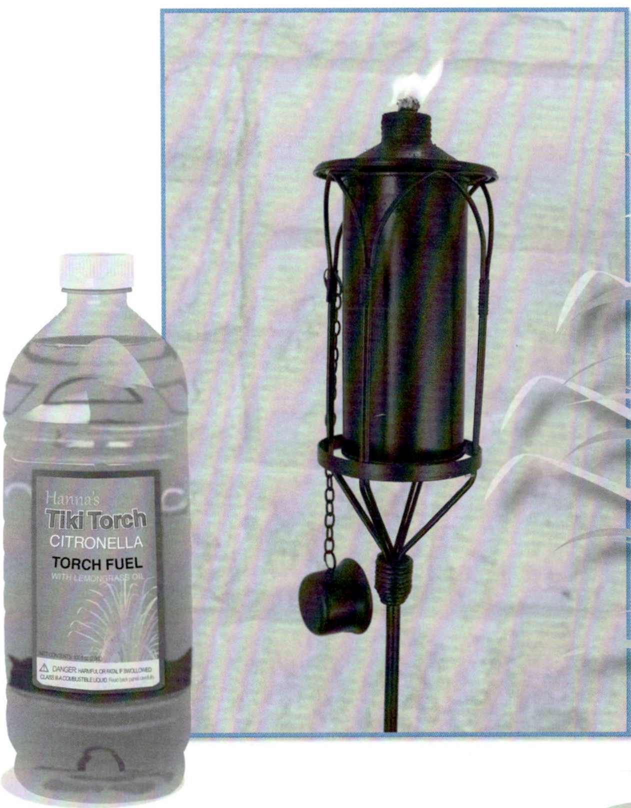

Photograph by Hanna Roach

© REGION 4 EDUCATION SERVICE CENTER. ALL RIGHTS RESERVED.

UNIT 1: Thinking like a Scientist

Lesson 2: Science and Scientists

Let's Explore!

Use research tools to find a science word that begins with each assigned letter of the alphabet.

Write the word next to the letter.

Let's Explain!

In your science alphabet, is there—

- A famous scientist? Why is the scientist famous?
- A scientific tool? What is it used for?
- A living organism? Where does it live? Is it endangered?
- An invention? Does this invention improve your daily life? How?

Let's Elaborate!

In your science notebook, write the following sentence stem.

What would life be like without ________________?

1. Choose a card.
2. Glue the card in the blank.
3. Record your answer to the question.
4. List a science concept that the inventor of the object most likely needed to know to create the object.

Let's Evaluate!

Use your knowledge of science and scientists to complete the assessment.

© REGION 4 EDUCATION SERVICE CENTER. ALL RIGHTS RESERVED.

Lesson 3: Observations and Inferences

Let's Engage!

Listen carefully to your teacher's instructions.

Let's Explore!

Observe the photograph.

Read the statements.

1. The students are at recess.
2. Grade 5 students use this classroom.
3. There are no students in the photograph of the classroom.
4. It is a warm, sunny day.
5. There are computers in the room.

Identify which statements are observations of the photograph.

Be prepared to justify your answers.

Let's Explain!

Making Inferences

You observed that—

There are no students in the photograph of the classroom. You used your sense of sight to make this observation. This is called direct evidence.

So—

Where are the students? Are they at recess? If they are at recess, does that mean the weather is nice outside?

Answers to these questions are called inferences. An inference is an explanation based on a combination of observations and experiences. If you see that there are no students in the classroom and you know it is nice and sunny outside, then you might think the students are at recess. You come to this conclusion because your teacher takes you outside when the weather is good but not when it is bad. Your past experience is indirect evidence that supports your inference.

© REGION 4 EDUCATION SERVICE CENTER. ALL RIGHTS RESERVED.

Let's Elaborate!

A scientist makes observations and inferences of all types of data.

Observe the line graph.

State Forest Animal Populations

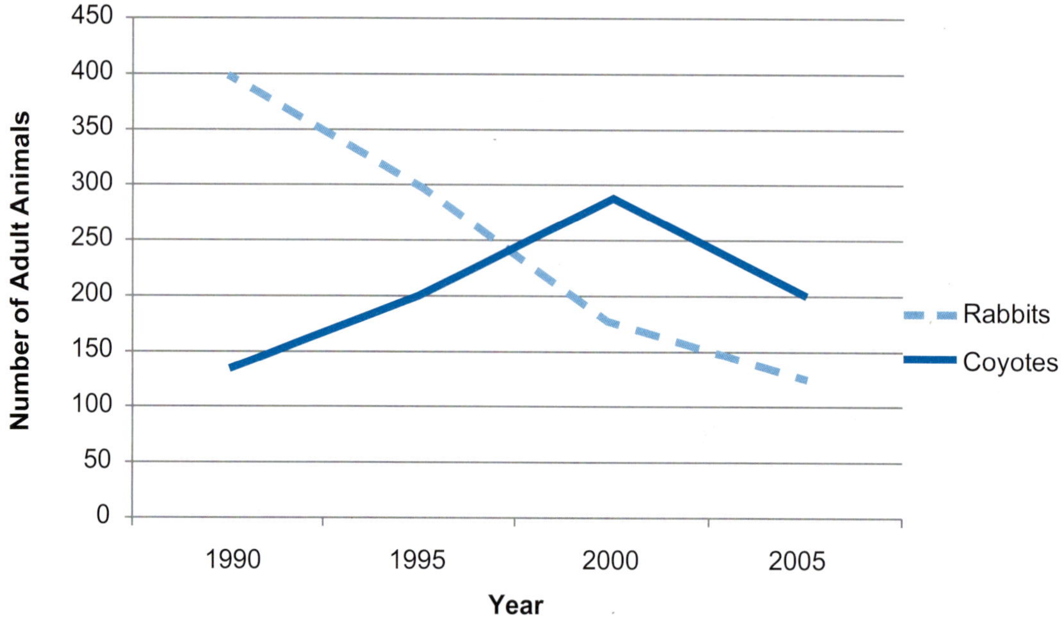

In your science notebook, copy the statements below and fill in the blanks with the correct information.

1. In 1990, there were__________adult rabbits in the state forest.
2. By 2000, the adult rabbit population had decreased to about ________________.
3. From 1990 to 2000, the coyote population________________.
4. From 1990 to 2000, the rabbit population________________.

Discuss and record your answers to the following questions.

1. Why do you think the rabbit population decreased when the coyote population increased?
2. After 2000, why do you think the coyote population began to decrease as the rabbit population decreased?
3. Which observations helped you make inferences?

Let's Evaluate!

Use your knowledge of observations and inferences to complete the assessment.

© REGION 4 EDUCATION SERVICE CENTER. ALL RIGHTS RESERVED.

Unit 2

Matter and Energy

UNIT 2: Matter and Energy

Introduction

Look around you. What do you see? Books? Desks? Plants? Your teacher? Other students? All these things are made of matter. In fact, everything that has mass and takes up space is made of matter.

How big is it? What color is it? What does it taste like? How does it feel? The answers to these questions are known as physical properties of matter. Physical properties of matter are observed by using your five senses. Some physical properties of matter may require using a tool such as a ruler, a thermometer, or a balance.

By the end of this unit, you will—

Know these concepts:

- Matter has mass and takes up space.
- Matter has physical properties that may be observed by the five senses and/or measured with tools.

Be able to do these things:

- Demonstrate safe practices in a science investigation.
- Make scientific observations and inferences.
- Collect and record scientific data.
- Make connections to prior knowledge.
- Make predictions based on observations.
- Communicate valid conclusions.
- Identify physical properties of matter.
- Classify matter based on physical properties.
- Measure length, mass, volume, and temperature.
- Use appropriate units with measurements.
- Identify mixtures and solutions.

Use these tools:

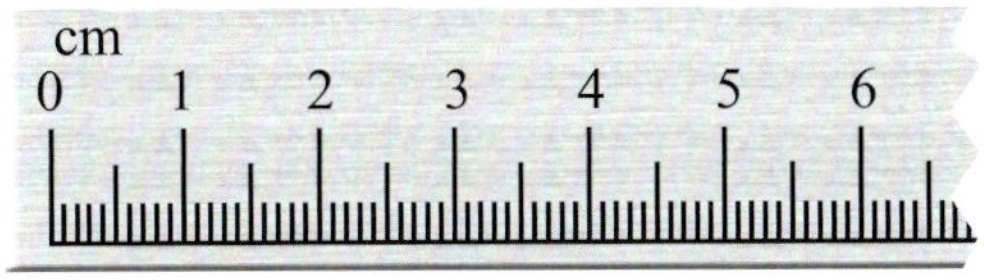

metric ruler
meter stick

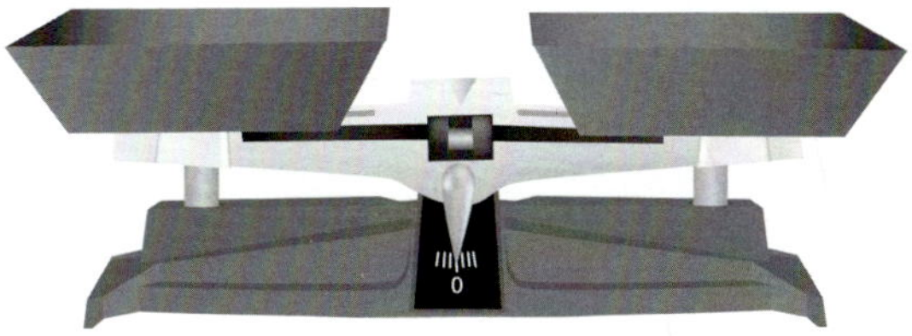

pan balance

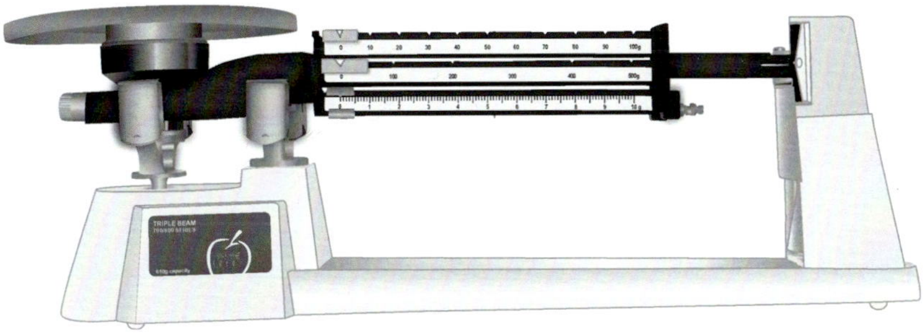

triple beam balance

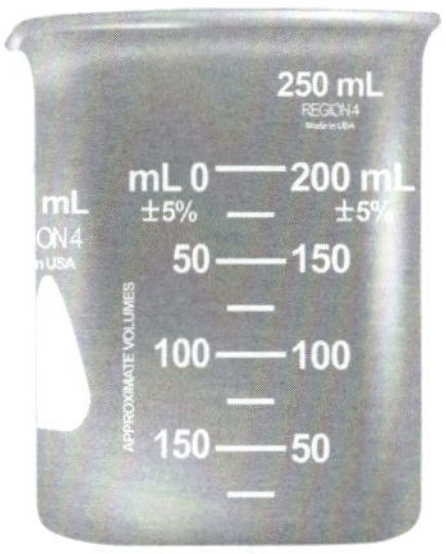

beaker

metric measuring cups
metric measuring spoons

calculator

graduated cylinder

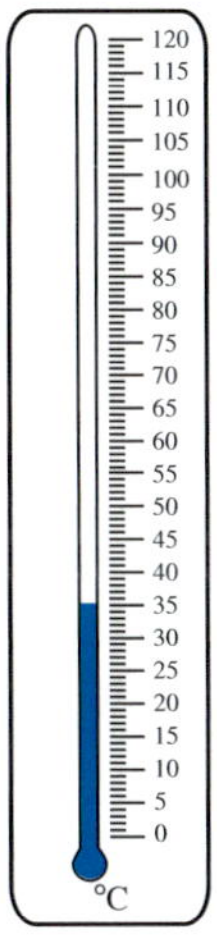

Celsius thermometer

magnets

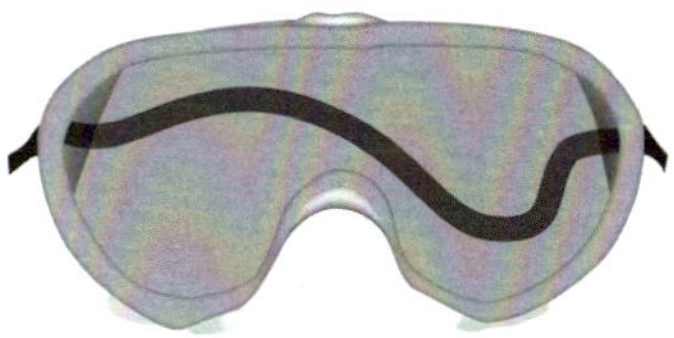

safety goggles

© REGION 4 EDUCATION SERVICE CENTER. ALL RIGHTS RESERVED.

UNIT 2: Matter and Energy

Lesson 1: Classifying Matter

Let's Engage!

Listen carefully to your teacher's instructions.

Let's Explore!

1. Observe the buttons.
2. Decide which physical property you will use to sort the buttons.
3. Classify the buttons.
4. Raise your hand when you are finished so your teacher can make observations of your work.

Record your answers to the following questions.

1. Which physical properties did you use to classify the buttons?
2. How many groups did you make?
3. How many buttons were in each group?

Let's Explain!

Our Senses

We use our senses to observe matter. A person can feel, taste, hear, see, and smell different types of matter. These observations describe the physical properties of matter.

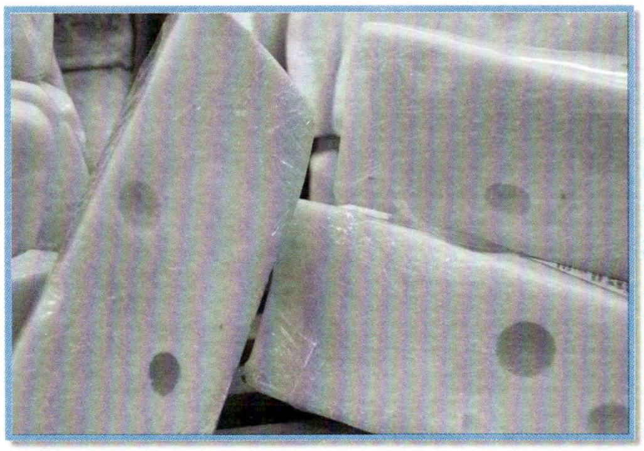

Photograph by Shelly D. Smith

For example, think about the physical properties of a piece of cheese.

What color is the cheese? How does it taste? How big is the piece of cheese?

Answers to these questions are physical properties of the cheese.

Physical properties may be used to classify matter. You used physical properties of the buttons to put them into groups.

Matter also can be classified by how it is used. The use of a material or object is called its function.

© REGION 4 EDUCATION SERVICE CENTER. ALL RIGHTS RESERVED.

UNIT 2: Matter and Energy

Lesson 1: Classifying Matter

Observe the following items.

- What is the function of each item?
- How can these items be classified by their use, or function?

1. Make a list of your 10 favorite foods.
2. Record at least three physical properties of each food on your list.
3. Classify your food choices using one of the physical properties of the foods.
4. Observe the concept map below created by a fifth-grade student.
5. Create your own concept map.

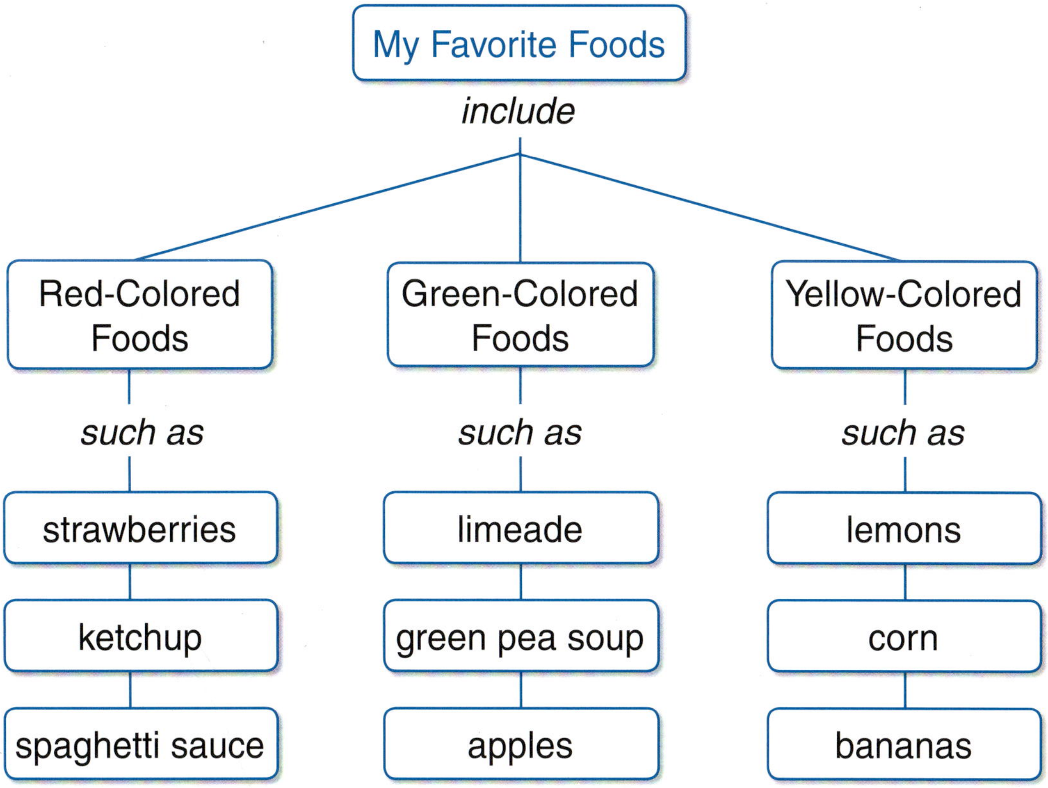

© REGION 4 EDUCATION SERVICE CENTER. ALL RIGHTS RESERVED.

Let's Elaborate!

Classification

Classification is used in our daily lives.

Students are classified by grade level at school.

The seats at a concert are grouped, or classified, by how much they cost. Tickets for seats closest to the stage cost more than tickets for seats in the back.

Groceries are classified by food types in a grocery store. For example, dairy foods, such as milk, butter, and yogurt, are displayed in the same area.

As a group—

- Identify another way that classification is used in our world.
- Create a table that shows how you could classify your information.

Let's Evaluate!

Use your knowledge of classifying matter to complete the assessment.

Let's Engage!

How tall are you?

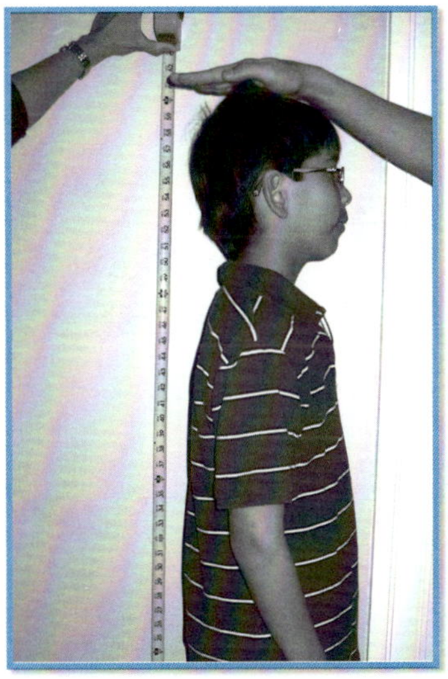

Photograph by Diana Elizondo

Let's Explore!

Scientists use the metric system to explore the world around us.

Observe the metric ruler.

- What does each number represent?
- What does each line represent?

Observe the meter stick.

- What does each number represent?
- What does each line represent?

Observe the measuring tape.

- What does each number represent?
- What does each line represent?

© REGION 4 EDUCATION SERVICE CENTER. ALL RIGHTS RESERVED.

Let's Explain!

Scientists measure physical properties of matter. It is very important for measurements to be accurate.

Observe the pencil.

- How long is the pencil in centimeters?

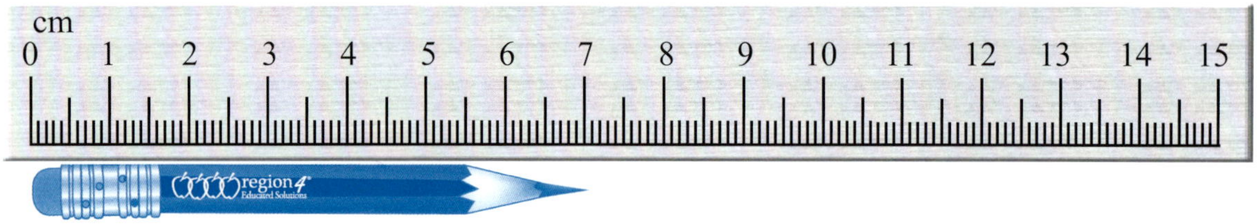

Observe the crayon.

- How long is the crayon in centimeters?
- How did you measure the crayon when it did not start at the zero mark?

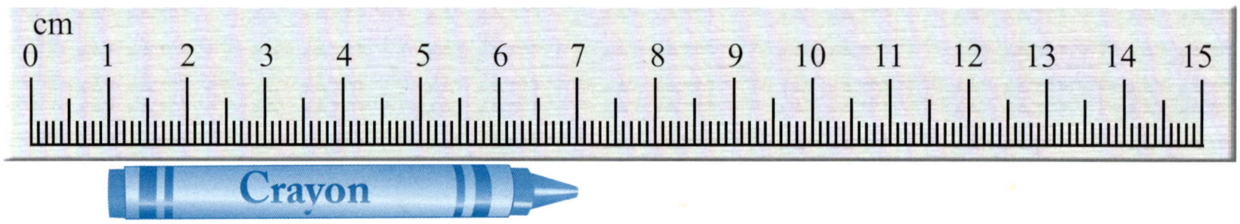

Millimeters (mm) are used to measure small objects.

Observe the ant.

- How long is the ant in millimeters?

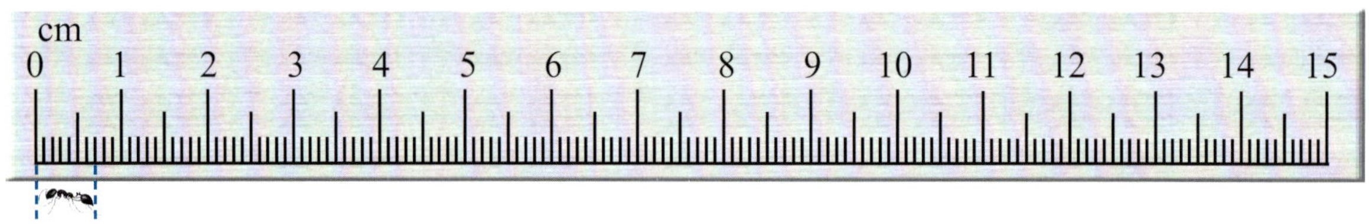

A kilometer is the unit of measurement for long distances, such as the distance between cities. A kilometer is equal to 1000 meters.

Copy the information into your science notebook and fill in the blanks.

1. 1 meter (m) = _______centimeters (cm)
2. 1 meter (m) = _______millimeters (mm)
3. 1 centimeter (cm) = _______millimeters (mm)
4. 1 kilometer (km) = _______meters (m)
5. Sequence the units of measurement (centimeter, millimeter, meter, and kilometer) from least to greatest.

Practice reading a ruler.

Observe the rulers and objects.

Record the length of the objects. Don't forget your unit of measurement.

1.

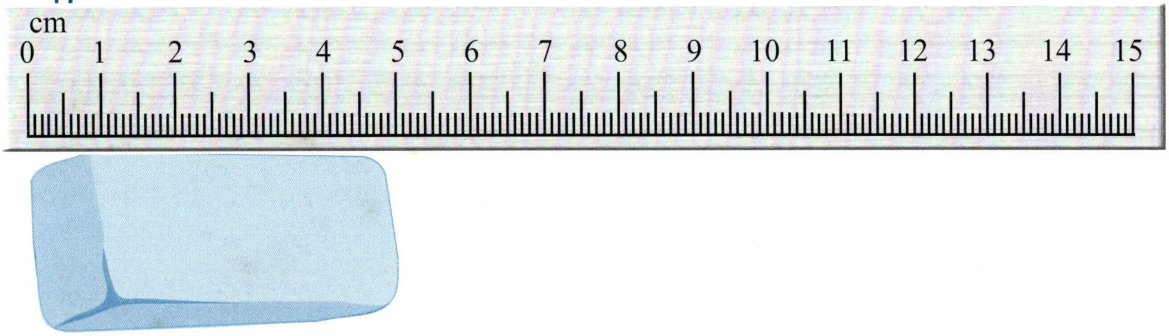

2.

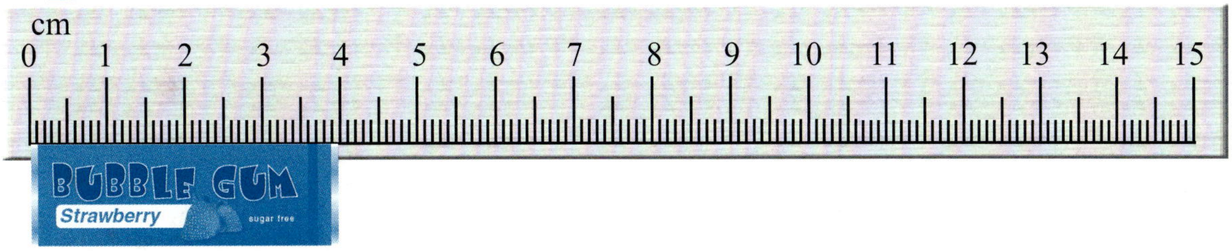

© REGION 4 EDUCATION SERVICE CENTER. ALL RIGHTS RESERVED.

3.

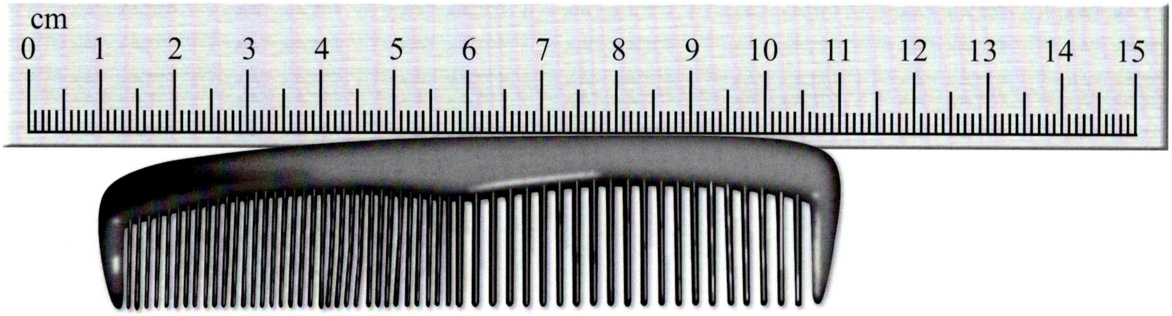

4.

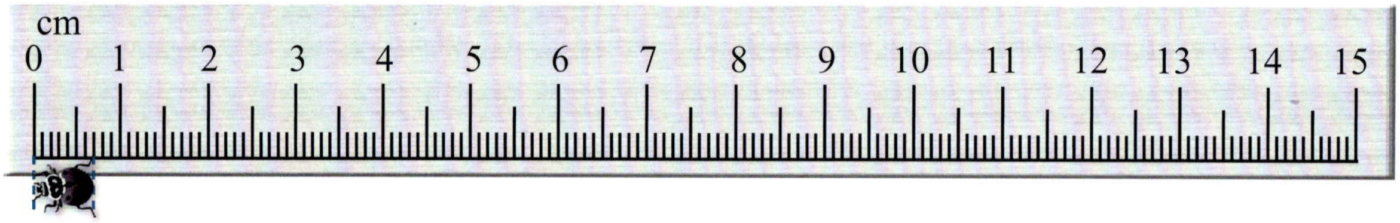

Let's Elaborate!

1. Read the questions below.
    - How tall is your classroom door?
    - What is the length of your favorite pencil?
    - How wide is your science book?
    - What is the distance between your desk and the door?
    - What is the distance around the top of your desk?
    - What is the length of your shoe?
2. Predict if each measurement is <50 cm or >50 cm.
3. Record your predictions in the table by placing a check mark in the appropriate box.
4. Test your predictions by choosing the best tool to make your measurements.
    - How close are your predictions?

Let's Evaluate!

Use your knowledge of measuring length to complete the assessment.

© REGION 4 EDUCATION SERVICE CENTER. ALL RIGHTS RESERVED.

UNIT 2: Matter and Energy

Lesson 3: Measuring Mass

Let's Engage!

Observe and compare the objects on your lab table. Be ready to share your observations.

Let's Explore!

Part 1

Pay close attention to the teacher demonstration.

Ask questions if you do not understand something.

Part 2

Copy the table as shown below.

Object	Predicted Mass in Grams (g)			

Follow the procedure on the activity card.

Let's Explain!

Mass

The amount of matter in an object is its mass.

Mass is a physical property that may be measured using a balance. The basic unit of measurement for mass is the gram (g). Large amounts of mass are measured in kilograms (kg). Moving an object from one location to another does not change the mass of an object.

pans

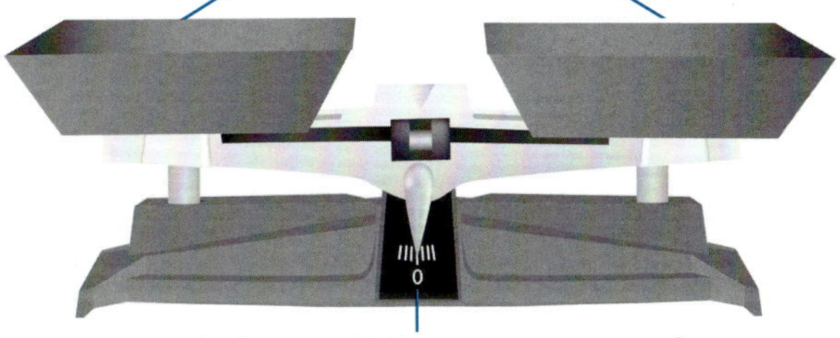

pointer, at the zero mark

pan balance

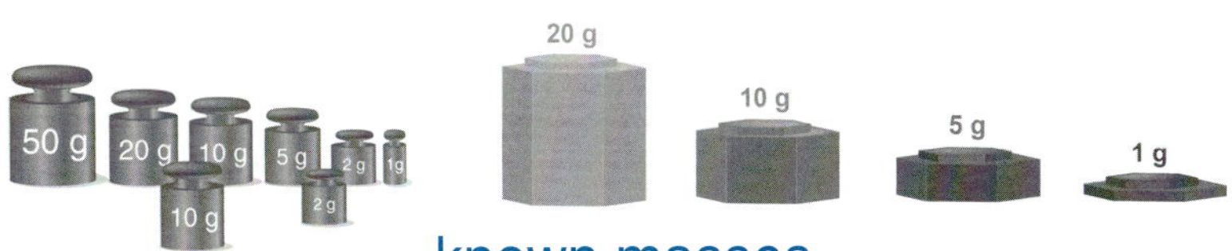

known masses

pan

riders

pointer, at the zero mark

adjustment screw

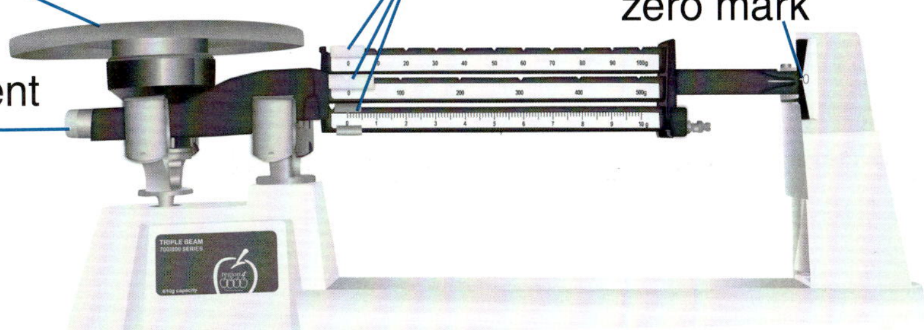

triple beam balance

© REGION 4 EDUCATION SERVICE CENTER. ALL RIGHTS RESERVED.

Let's Elaborate!

Part 1: Testing Predictions

1. Fill in the headings of the remaining columns in your table.
2. Test your predictions using a balance.
3. Calculate the difference between your prediction and the actual measurement.

Object	Predicted Mass in Grams (g)	Actual Mass in Grams (g)		Difference in Grams (g)
		Pan	Triple Beam	

Part 2: Measuring Liquid in a Container

1. Using the materials at your lab table, measure the mass of the liquid in the container.

   Think about—

   - Did you measure the mass of the liquid? Or
   - Did you measure the mass of the liquid and the container?

2. Repeat the activity if necessary.
3. Be ready to demonstrate how your group measured the mass of the liquid in the container.

Part 3: Calculating the Mass of a Liquid

Observe the triple beam balance below.

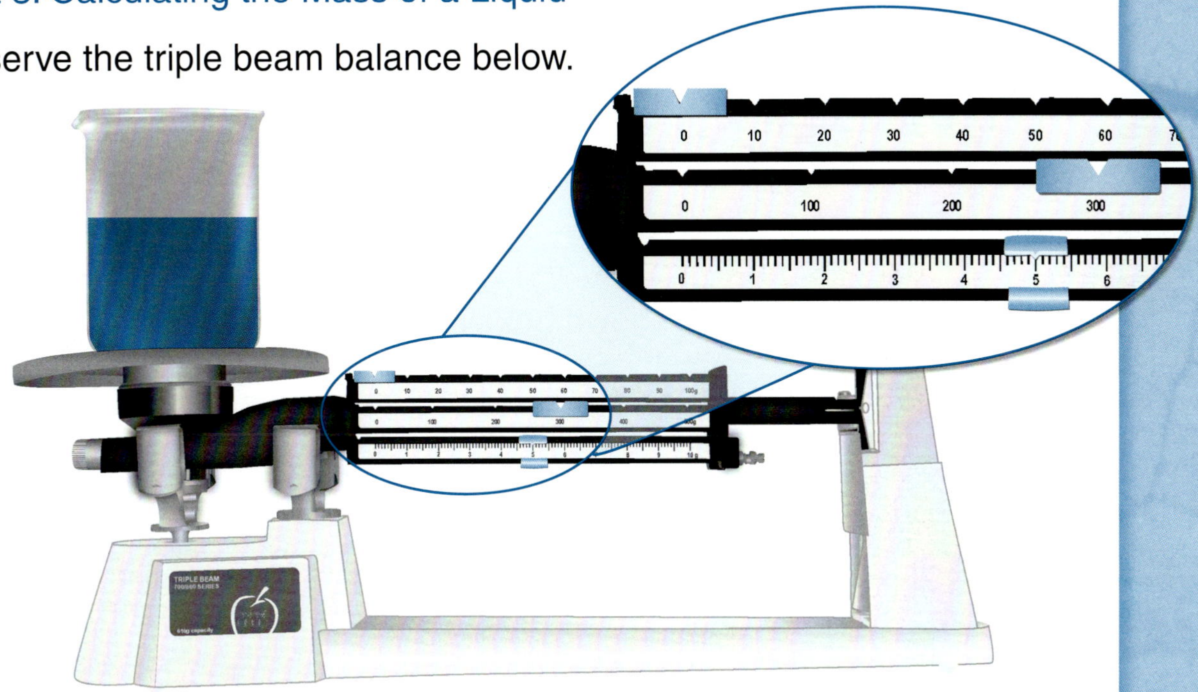

To measure the mass of a liquid, the liquid must be poured into a container.

The mass of the beaker and the liquid is 305 grams.

How do you measure the mass of the liquid in a container?

1. Measure the mass of the empty container. In this example, the beaker has a mass of 105 grams.
2. Pour the liquid you are measuring into the container.
3. Measure the mass of the container and its contents.
4. Read the measurement on the balance. In this example, the measurement is 305 grams.
5. Subtract the mass of the empty container (105 g) from the mass of the container and its contents.

305 g – 105 g = 200 g

mass of beaker of liquid — mass of empty beaker — mass of liquid

© REGION 4 EDUCATION SERVICE CENTER. ALL RIGHTS RESERVED.

UNIT 2: Matter and Energy

Lesson 3: Measuring Mass

Let's Evaluate!

Use your knowledge of measuring mass to complete the assessment.

Part 1

1. Draw the T-chart shown below on chart paper.

Mass <50 Grams	Mass >50 Grams

2. Observe the items provided to your group.
3. Measure the mass of each object using a balance.
4. Classify the objects in the T-chart.
5. Record the name of each object in the correct column.

Part 2

Follow your teacher's instructions.

Let's Engage!

Identify—

- the tool
- the physical property being measured
- the measurement
- the unit of measurement

1. 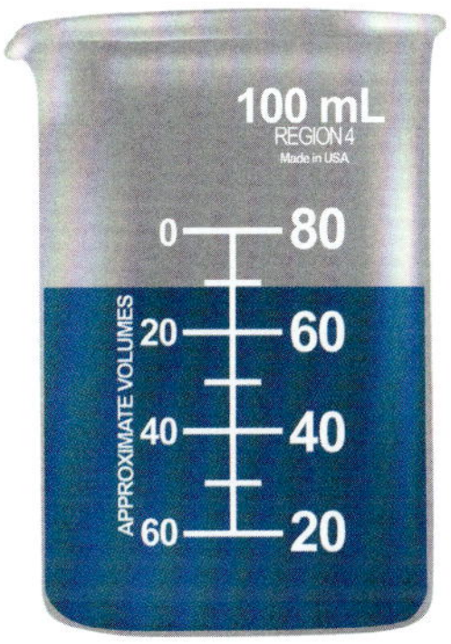

2. 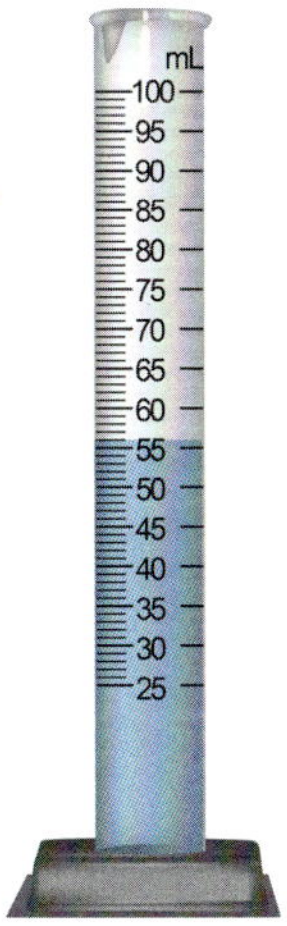

3. 

Let's Explore!

1. Observe the containers on the table.
    - Which container has the greater volume?
2. Plan an investigation to test your prediction. Think about the following questions as you plan the investigation:
    - What materials are needed?
    - Does the investigation actually test your prediction?
3. In your science notebook, write the steps of the investigation.
4. Ask your teacher to review and approve your plan.

© REGION 4 EDUCATION SERVICE CENTER. ALL RIGHTS RESERVED.

5. Gather the remaining containers from the display table and carry out the investigation to test your prediction.
    - Which container has the greatest amount of space inside?
    - Does the result of your investigation match your prediction?

Let's Explain!

Planning an Investigation

A scientist makes observations.

- What observations of the containers did you make?

Making observations can lead to asking questions.

- What question did you investigate?

A scientist plans investigations.

- How did you plan your investigation?
- How did you decide which containers to use?

An investigation can guide a scientist to find answers to questions.

- Did your investigation actually test your prediction?
- What were the results of the investigation?

Let's Elaborate!

Volume is a physical property of matter. Mass is a physical property of matter.

The volume and mass of water are related in a very special way in science.

1. Follow the procedure on the activity card.
2. Copy the following sentence into your science notebook:

   One mL of pure water has a mass of 1 gram.

3. Imagine filling up a centimeter cube (1 cm^3) with water.

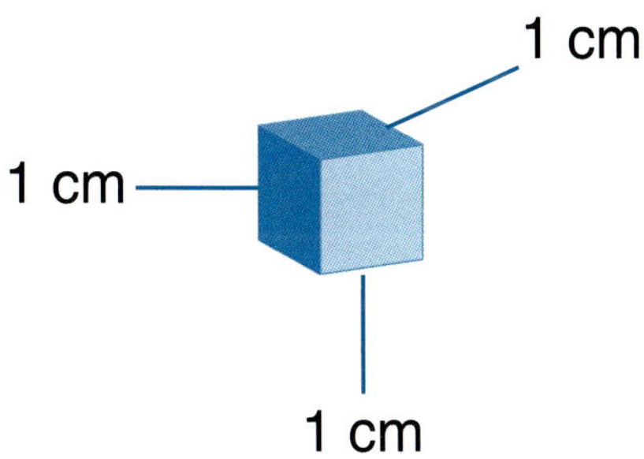

One milliliter (1 mL) of pure water fills 1 cm^3.

Let's put it all together.

One mL of pure water has a mass of 1 gram and a volume of 1 cm^3.

Let's Evaluate!

Use your knowledge of measuring volume to complete the assessment.

© REGION 4 EDUCATION SERVICE CENTER. ALL RIGHTS RESERVED.

Let's Engage!

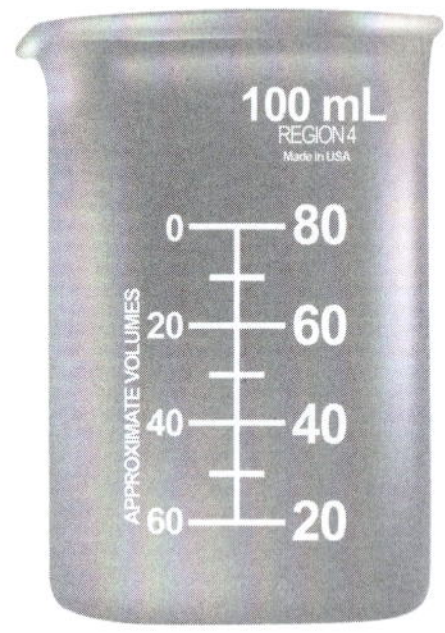

Observe the demonstration.

Think about—

- What liquids are poured into the beaker?
- What are the physical properties of those liquids?
- Do the physical properties of the liquids change when they are combined?

Let's Explore!

Copy the table into your science notebook.

Container	Mass (g)	Volume (mL)
A (cotton balls)		
B (gravel)		

1. Make observations of the two small plastic containers.
2. Measure and record the volume of Container A using the materials provided. Since the containers are identical, record the same volume for Container B.
3. Fill Container A with as many cotton balls as possible. Close the lid tightly.
4. Measure the mass of Container A and its contents. Record the mass in the table.
5. Fill Container B with as much gravel as you can. Close the lid tightly.
6. Measure the mass of Container B and its contents. Record the mass in the table.
   - Which container of matter has the greater mass? How do you know?
   - Does one container of matter take up a greater amount of space than the other? How do you know?
   - Which container has matter that is more tightly packed together? How do you know?

© REGION 4 EDUCATION SERVICE CENTER. ALL RIGHTS RESERVED.

Let's Explain!

Measuring Density

Container A
(cotton balls)

Container B
(gravel)

Container A and Container B are exactly the same size and shape. They have the same volume.

When the containers are filled with the cotton balls and gravel, Container B has more mass than Container A. Density describes how much matter is packed into a given amount of space. Container B has more mass packed into its volume than Container A. Container B has a greater density than Container A.

What about the oil and water mixture? The oil floats on top of the water because the oil is less dense than the water. More matter is packed into the same volume of water than in the same volume of oil.

Let's Elaborate!

You have observed physical properties of density cubes. The cubes are different colors, but they are the same size and shape.

Part 1

Color of Cube	Mass (g)	Volume (cm^3)	Sinker or Floater?

1. Copy the table into your science notebook.
2. Using the materials at your station, measure and record in the table the mass of each density cube.
3. Do not record data in the other columns at this time.

You have measured the mass of each density cube. Each cube has a different mass.

© REGION 4 EDUCATION SERVICE CENTER. ALL RIGHTS RESERVED.

Part 2

Measuring Volume

Volume is the amount of space that matter takes up. The density cubes are the same size and shape. The volume of each density cube is the same.

What is the actual volume of each density cube? How is it measured? Observe the demonstration closely.

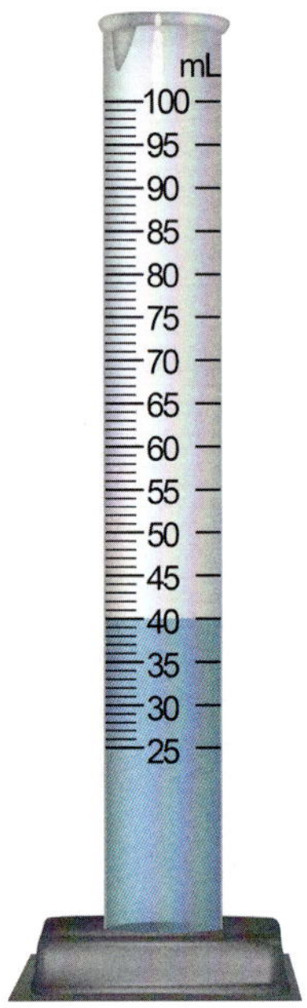

water without cube
(40 mL)

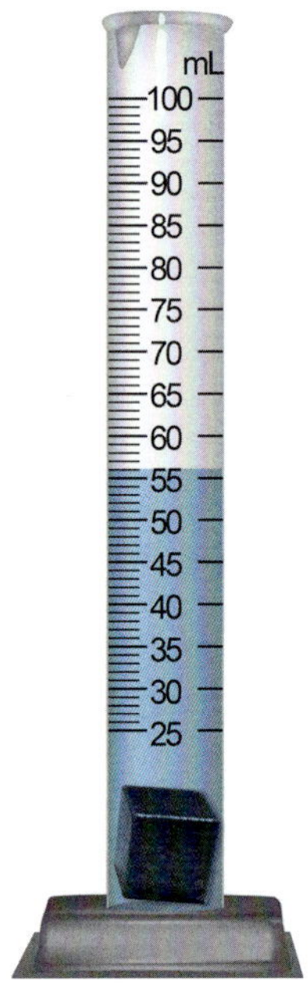

water with cube
(56 mL)

The level of the water rises 16 mL when the cube is placed in the water. The difference in the two water levels is equal to the volume of the cube.

56 mL – 40 mL = 16 mL

The level of water increased 16 mL. Since 1 mL = 1 cm^3, the volume of the cube is equal to 16 cm^3.

What if the cube floats?

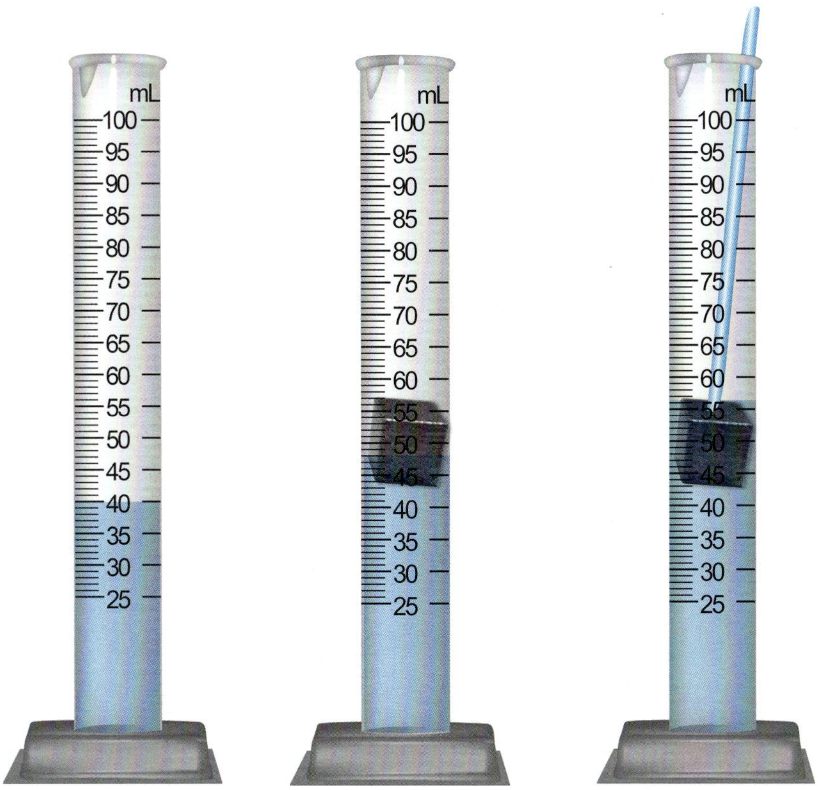

If the cube floats, you may take a pencil or a straw and gently push on the cube until it is completely underwater. This process of measuring the volume of an object is called water displacement.

Practice—

Use water displacement to find the volume of the marble and the small rock. Be prepared to model this process for your teacher.

Part 3

Which density cubes float? Which density cubes sink?

1. Place the density cubes, one at a time, in a container of water.
2. Record if each cube is a sinker or a floater.
3. Using a highlighter, draw a line through every cube that is a sinker.

© REGION 4 EDUCATION SERVICE CENTER. ALL RIGHTS RESERVED.

4. Using a different color highlighter, draw a line through every cube that is a floater.

5. Analyze your data.

Think about—

- Which physical property of each sinker was greater, mass or volume?
- Which physical property of each floater was greater, mass or volume?
- What valid conclusions can be made about the density cubes?

If the mass (g) of the density cube is greater than its volume (cm^3), it sinks in water. The density of the cube is greater than the density of the water.

If the mass (g) of the density cube is less than its volume (cm^3), it floats in water. The density of the cube is less than the density of the water.

Record the following:

Mass (g) $>$ Volume (cm^3): the object sinks in water

Mass (g) $<$ Volume (cm^3): the object floats in water

Let's Evaluate!

Use your knowledge of density to complete the assessment.

1. Copy the table into your science notebook. Use it to record your data.

Object	Prediction: Sinker or Floater?	Mass (g)	Volume (cm^3)	Actual: Sinker or Floater?

2. Record the names of the objects in the table.
3. Predict and record whether each object is a sinker or a floater.
4. Measure and record the mass of each object.
5. Measure and record the volume of each object.
6. Test your predictions.
7. Record in the final column whether the object is a sinker or floater.
8. Using your knowledge of physical properties, explain why objects float or sink in water.

© REGION 4 EDUCATION SERVICE CENTER. ALL RIGHTS RESERVED.

Let's Engage!

Examples of Physical Properties

color
shape
odor
texture
mass
temperature
volume

1. Discuss the physical properties of each item on the list.

- apple
- milk
- oxygen
- rock
- baseball
- helium
- ice
- juice
- marble
- nitrogen
- water vapor
- pencil
- water
- gasoline

2. Copy the table below.

Column 1	Column 2	Column 3

3. Fill in the table as instructed by your teacher.

Let's Explore!

Stations

1 Air

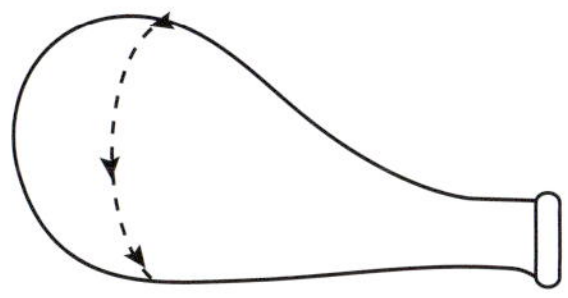

2 Water

3 Marbles

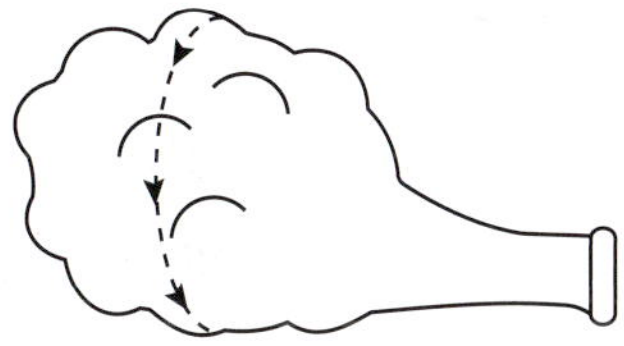

1. Rotate through each station.
2. Follow the classroom safety rules.
3. Read the lab card before beginning the investigation.
4. Follow the procedure step by step.
5. Make accurate measurements.
6. Ask questions.
7. Make observations.

L ROTATION

SAFETY ALERT

Report spills immediately.

© REGION 4 EDUCATION SERVICE CENTER. ALL RIGHTS RESERVED.

Let's Explain!

The Three States of Matter

Physical properties of matter include shape, mass, and volume. The state of matter is also a physical property. There are three states of matter: solid, liquid, and gas.

Matter is made of very small particles that are constantly moving and bumping into each other. The particles in each state of matter act very differently.

Particles in a solid state of matter are packed tightly together and move very little, nearly staying in place. That is one reason a solid keeps its shape when moved from place to place.

In a liquid, the particles are more spread out, move faster than in a solid, and flow around each other. A liquid flows as it is poured from one container to another. It takes the shape of the new container every time.

In a gas, the particles are spread far apart and move quickly. A gas spreads out to fill its container.

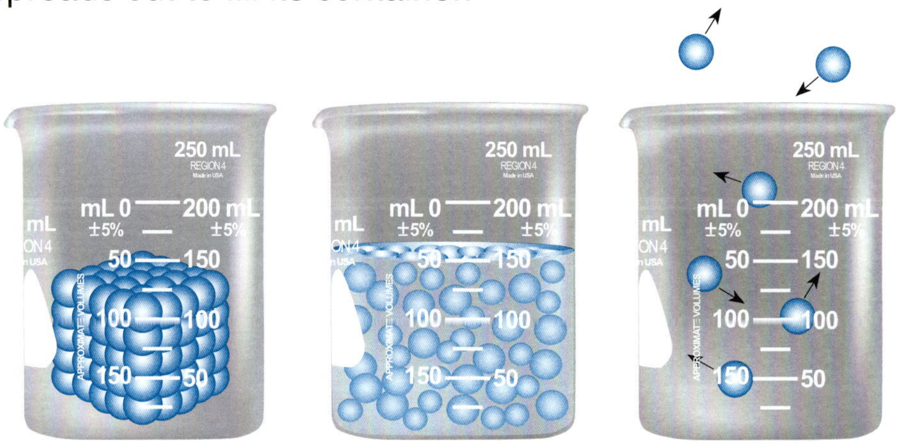

solid liquid gas

1. Copy the models into your science notebook and label them.

2. Define each state of matter in your own words.

Apple

When an apple is moved from one container to another, it keeps its shape. The space it takes up, or its volume, stays the same. An apple is a solid. Other examples of solids are a rock, a baseball, and a pencil.

Milk

When milk is poured from one container to another, it takes the shape of the container. The space it takes up, or its volume, stays the same. Milk is a liquid. Cooking oil and juice are also liquids.

Balloon

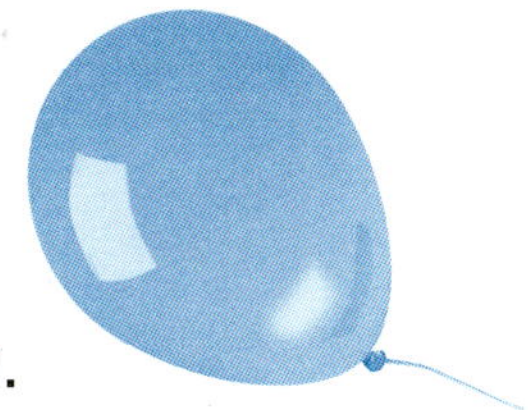

When an inflated balloon is popped, the air inside the balloon spreads out into the room. The space the air takes up, or its volume, changes. Air is a gas. Other examples of gases include oxygen, nitrogen, and helium.

Not all matter is classified so easily.

What about sand? Sand flows when it is poured from one container to another. It takes the shape of its container. Is sand a liquid? No, sand is a solid.

When classifying matter, make observations of an individual particle. A sand particle does not change its shape when moved from one location to another. Sand is a solid.

Let's Elaborate!

Part 1

1. Which of the following statements are reasonable and valid conclusions of the investigation of air? Justify your answer.

   a. Air has mass and takes up space.

   b. The state of matter of air is gaseous.

© REGION 4 EDUCATION SERVICE CENTER. ALL RIGHTS RESERVED.

2. Which of the following statements are reasonable and valid conclusions of the investigation of water? Justify your answer.

    a. Water has mass and takes up space.

    b. The state of matter of water is liquid.

3. Which of the following statements are reasonable and valid conclusions of the investigation of the marble? Justify your answer.

    a. The marble has mass and takes up space.

    b. The state of matter of a marble is solid.

Part 2

1. Make and share observations of the materials at your table.
2. Place the effervescent tablet into the cup of water.
3. Make, share, and record observations.
4. Take turns placing a hand above the cup.
5. Make, share, and record observations.
6. Draw the setup.
7. Label the three states of matter that can be observed in the setup.

Let's Evaluate!

Use your knowledge of the states of matter to complete the assessment.

© REGION 4 EDUCATION SERVICE CENTER. ALL RIGHTS RESERVED.

Lesson 7: Boiling and Freezing/Melting Points

Let's Engage!

Make and carry out a simple plan to melt the ice cube. Be ready to share and demonstrate your plan with the other groups.

Let's Explore!

1. Follow the classroom safety rules.
2. Read the lab card before beginning the investigation.
3. Follow the instructions carefully.
4. Make and record accurate observations.

Discuss and record answers to the following questions:

1. What is the material in each cup?
2. What are the physical properties of each type of matter?
3. Do any of the physical properties change beneath the heat lamp?
4. What are those changes?

Let's Explain!

Measuring Temperature

At what temperature does each type of matter change from a liquid to a solid?

What tool is used to measure temperature?

A thermometer measures temperature. Temperature is measured in units called degrees. Scientists measure temperature in degrees Celsius (°C). For example, the temperature in your classroom is most likely about 25 °C. Water freezes at 0 °C, and it boils at 100 °C.

Knowing how to read the temperature on a thermometer is important.

Think about—

- A student wants to make fudge. The recipe says to heat the milk to 85 °C. If the milk is not heated correctly, the fudge may be too thin or too hard.
- A student is hatching eggs in an incubator for a science investigation. The instructions are to keep the eggs at 38 °C. If the incubator is too hot or too cold, the eggs may not hatch.

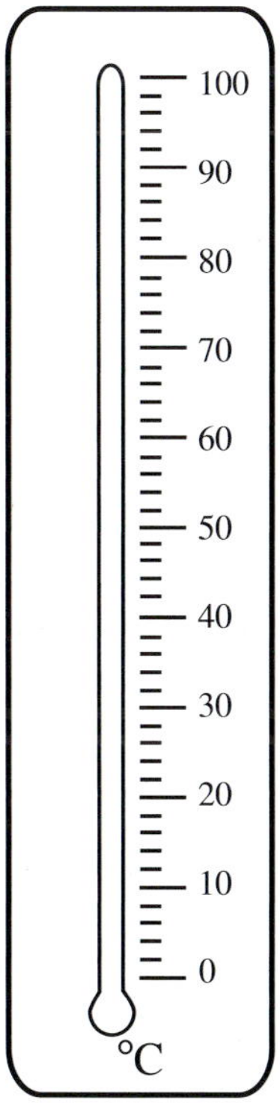

You already know that it is important to know what the lines on a measuring tool represent. The same is true for a thermometer.

Observe the pictures of the thermometers.

The numbers on this thermometer are 10 degrees apart. Each line represents 2 degrees.

© REGION 4 EDUCATION SERVICE CENTER. ALL RIGHTS RESERVED.

UNIT 2: Matter and Energy

Lesson 7: Boiling and Freezing/Melting Points

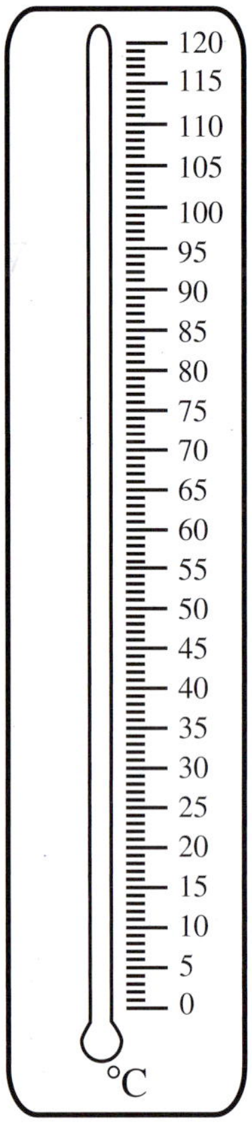

The numbers on this thermometer are 5 degrees apart. Each line represents 1 degree.

The liquid inside the thermometer goes up as the temperature increases, and it goes down as the temperature decreases.

Some thermometers are digital. There is no column of liquid that moves up and down in this type of thermometer. The temperature reading is simply shown on a digital display.

Lesson 7: Boiling and Freezing/Melting Points

Record the measurement of temperatures shown on each thermometer.

1.

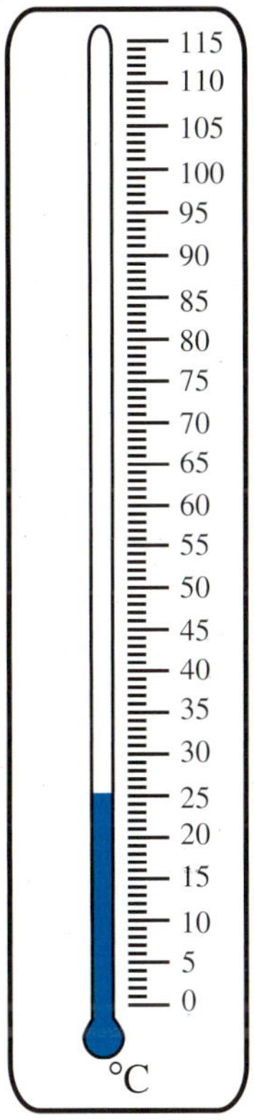

2.

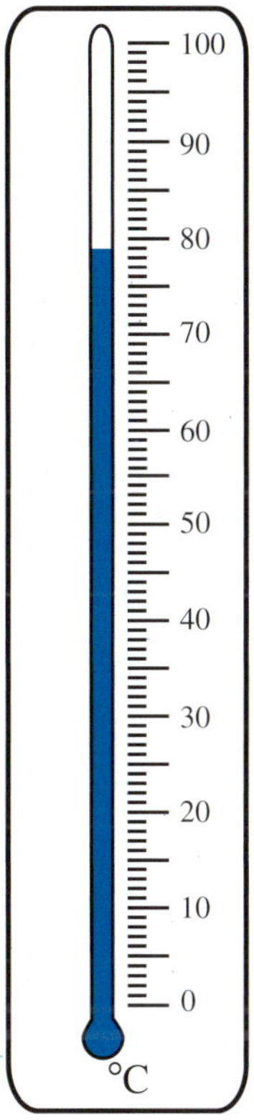

3.

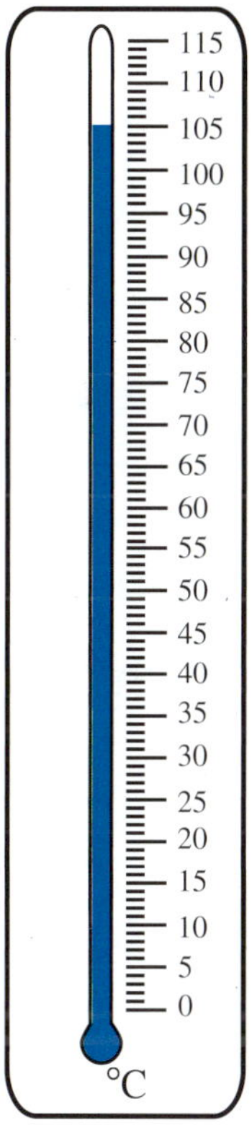

4.

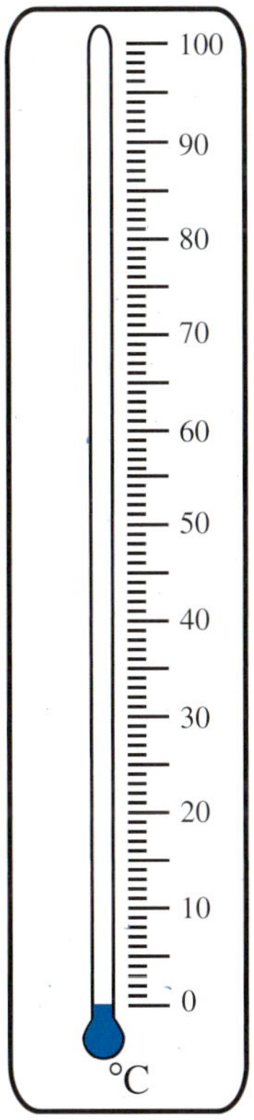

5.

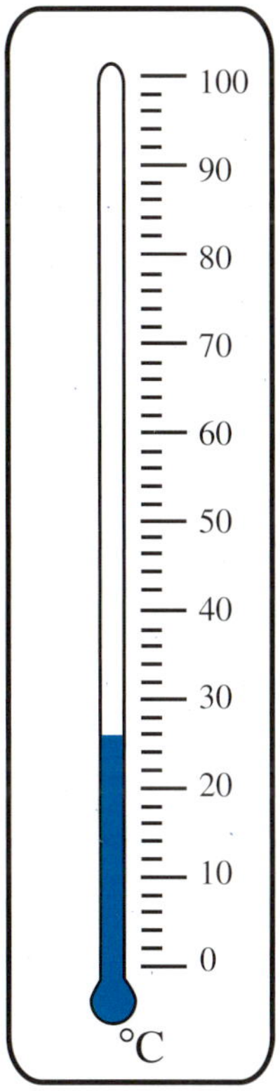

6.

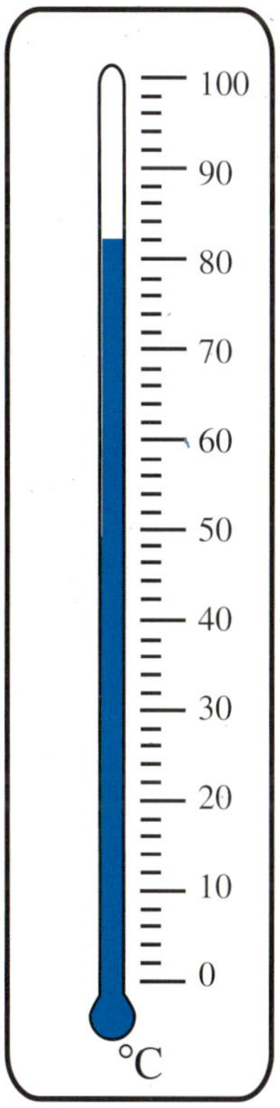

© REGION 4 EDUCATION SERVICE CENTER. ALL RIGHTS RESERVED.

Discuss and record answers to the following questions:

7. Which thermometer(s) display a temperature at which water is a liquid?

8. Which thermometer(s) display a temperature at which water begins to change from a liquid to a solid?

Let's Elaborate!

Melting Points and Boiling Points

You know that—

- Matter is most often found in three states: solid, liquid, and gas.
- States of matter may change when heat is added or reduced.

Think about water.

Water changes its state of matter from a liquid to a solid when the temperature falls to 0 °C. This temperature is the freezing point of water.

An ice cube changes its state of matter from a solid to a liquid when the temperature rises to 0 °C. This temperature is the melting point of the ice cube.

Water becomes water vapor when the temperature rises to 100 °C. At this temperature, the state of matter of water changes from a liquid to a gas. This temperature is the boiling point of water.

Lesson 7: Boiling and Freezing/Melting Points

Copy the table into your science notebook.

Table 7.1. Changes in State of Matter

Process	Changes in State of Matter
freezing	liquid to solid
melting	solid to liquid
boiling	liquid to gas

Not all substances have the same melting points or boiling points.

Analyze the data in the table below.

Table 7.2. Melting Points and Boiling Points

Type of Matter	State of Matter at Room Temperature	Melting Point	Boiling Point
water	liquid	0 °C	100 °C
salt	solid	801 °C	1465 °C
gold	solid	1064 °C	2807 °C
copper	solid	1083 °C	2595 °C
iron	solid	1535 °C	3000 °C

Use Table 7.2. Melting Points and Boiling Points to answer the following questions.

1. If the temperature of water is 110 °C, is it a liquid or a gas? Why?
2. If the temperature of gold is 1000 °C, is it a liquid or a solid? Why?
3. If the temperature of a piece of copper is 2000 °C, is it a liquid or a solid? Why?

© REGION 4 EDUCATION SERVICE CENTER. ALL RIGHTS RESERVED.

Let's Evaluate!

Use your knowledge of boiling and freezing/melting points to complete the assessment.

Let's Engage!

Why are cold items kept in an ice chest on a picnic?

Why does a chef use a potholder to remove a cake from the oven?

Why are the electrical wires on a lamp wrapped?

Let's Explore!

Follow the procedure on the activity card.

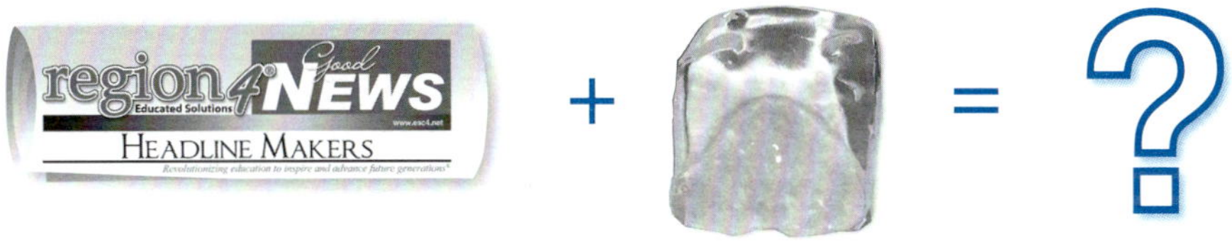

© REGION 4 EDUCATION SERVICE CENTER. ALL RIGHTS RESERVED.

Let's Explain!

Conductors and Insulators

A material or substance that transfers energy easily is called a conductor. For example, aluminum and copper are good conductors of thermal energy and electrical energy.

A substance that does not transfer energy easily is called an insulator. Wood and plastic are examples of insulators.

Think about—

Most cooking pots are made from some type of metal. Metals are good conductors of heat.

Some pots have metal handles, and a potholder must be used to protect your hands from the heat. Potholders are often made from many layers of cloth. A potholder does not allow thermal energy to be transferred from the metal handle to your hand. The cloth is an insulator.

In Explore, which material is an insulator? Which material is a conductor?

Label the insulator and conductor in your drawing of the setup.

Let's Elaborate!

Observe the flashlight.

A flashlight is a system. The parts of the flashlight work together to perform the function of the flashlight.

The function of the flashlight is to provide light. Electrical energy must travel in a closed path from the battery to light the bulb.

Follow the instructions on the activity card and complete the investigation as a group.

Be ready to share your observations.

Let's Evaluate!

Use your knowledge of conductors and insulators to complete the assessment.

© REGION 4 EDUCATION SERVICE CENTER. ALL RIGHTS RESERVED.

UNIT 2: Matter and Energy

Lesson 9: Mixtures and Solutions

Let's Engage!

Be ready to answer questions about what you observe.

Let's Explore!

Stations

1 Party Time

2 Tea for Two

3 Clips and Marbles

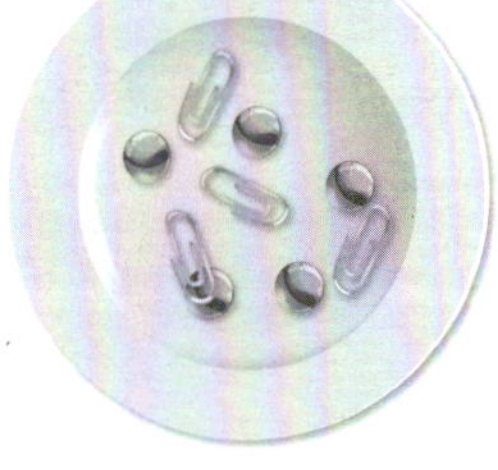

4 Surf's Up

As you rotate with your group through the stations, remember to—

1. Follow the classroom safety rules.
2. Read the lab card before beginning the investigation.
3. Follow the instructions.
4. Record data in your science notebook.
5. Make accurate measurements.
6. Ask questions.
7. Share observations.

Let's Explain!

Mixtures

A mixture is a combination of two or more things.

Think about—

Combining sliced bananas, sliced peaches, orange sections, and whole green grapes can make a fruit salad. This is a mixture. Each fruit in the salad keeps its own physical properties. For example, the sliced bananas still taste like a banana. The grapes are still round. The parts of a mixture can be separated by physical means. In the fruit salad, each fruit can be scooped out with a spoon and separated from the others.

Some mixtures can be separated using the physical property of magnetism. A magnet attracts substances that contain iron. Other tools that may be used to separate mixtures are filters, strainers, or even your fingers.

© REGION 4 EDUCATION SERVICE CENTER. ALL RIGHTS RESERVED.

A solution is a special type of mixture. In a solution, one substance is dissolved in another substance. The ability of a substance to dissolve in another substance is called solubility. Solubility is a physical property of matter.

Combinations of liquids, solids, or gases may form solutions.

Think about—

A lemonade mix dissolves in water to form lemonade. The lemonade is a solution.

The air we breathe is a solution. It contains oxygen dissolved in nitrogen.

Combining melted copper and tin forms a metal called bronze. Bronze is a solution.

Sugar is soluble in brewed coffee. The mixture of sugar and brewed coffee is a solution.

One way that the ingredients in solutions may be separated is by the process of evaporation. Evaporation occurs when the state of matter changes from a liquid to a gas. Energy is necessary for evaporation to happen.

You observed the results of evaporation in the Petri dish from "Tea for Two." The water in the tea solution evaporated into the air. The powdered instant tea was left in the Petri dish. The source of energy was the heat in the air.

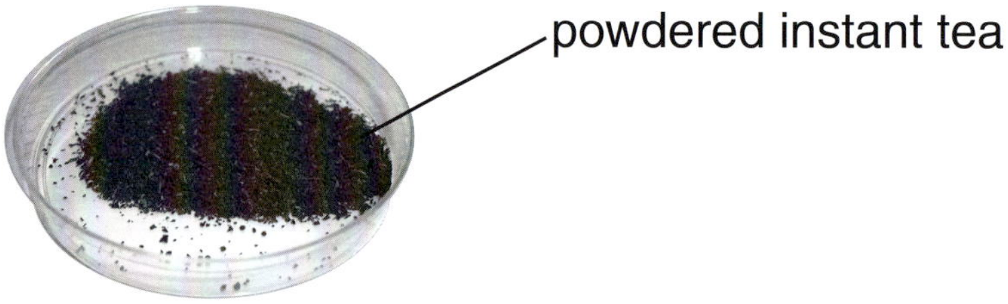

Evaporation can also be observed when water vapor rises above a pot of boiling water. The boiling point of water is 100 °C. The state of matter of water changes from a liquid to a gas. The source of energy is the hot plate.

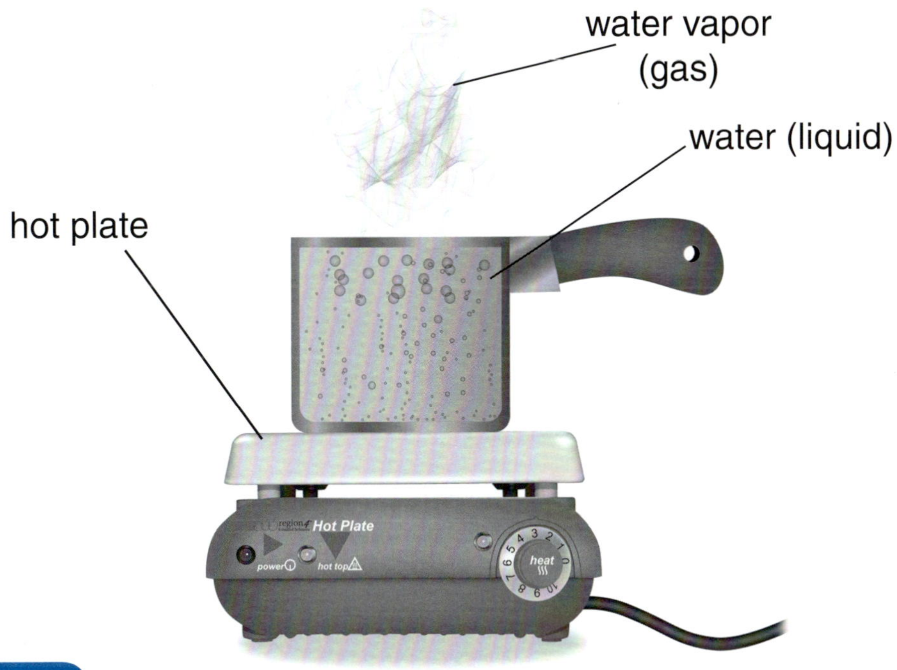

Let's Elaborate!

During a science investigation, a student combines 100 mL of white sand and 100 mL of salt in a resealable plastic bag and shakes it up.

Use your knowledge of mixtures and solutions to design, plan, and implement an experimental investigation to answer the question:

- How can the student completely separate the white sand and the salt?

Let's Evaluate!

Use your knowledge of mixtures and solutions to complete the assessment.

© REGION 4 EDUCATION SERVICE CENTER. ALL RIGHTS RESERVED.

Unit 3

Force, Motion, and Energy

UNIT 3: Force, Motion, and Energy

Introduction

Have you ever kicked a ball, picked up a backpack, or pulled a rope?

Each of these actions requires force. Force causes movement.

Do you like to listen to music? Turn on a light when it is dark? Eat a hot meal? Energy is necessary for these things to happen. Energy occurs in many forms—mechanical, light, thermal, electrical, and sound energy.

By the end of this unit, you will—

Know these concepts:

- Force is a push or a pull that can affect the movement of objects.
- Force can be measured.
- Friction affects the movement of objects.
- Electrical energy flows through a circuit and can produce light, heat, and sound.
- Light energy moves in a straight line until it strikes an object or is transmitted through different mediums.
- Light can be reflected and refracted.

Be able to do these things:

- Demonstrate safe practices in a science investigation.
- Design an experiment.
- Make scientific observations.
- Collect and record scientific data.
- Make connections to prior knowledge.
- Make predictions based on observations.
- Communicate valid conclusions.
- Recognize different forms of energy.
- Demonstrate reflection and refraction of light energy.

Use these tools:

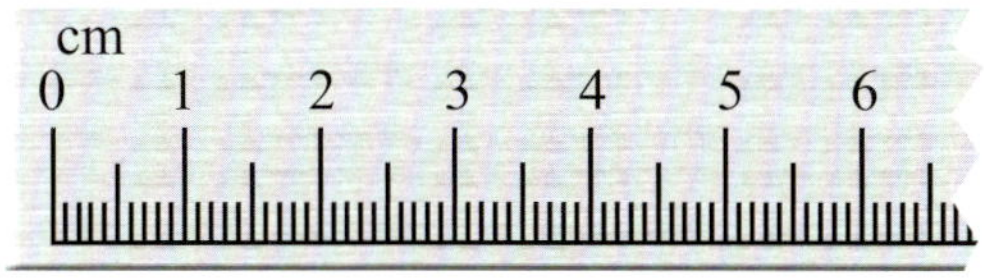

metric ruler
meter stick

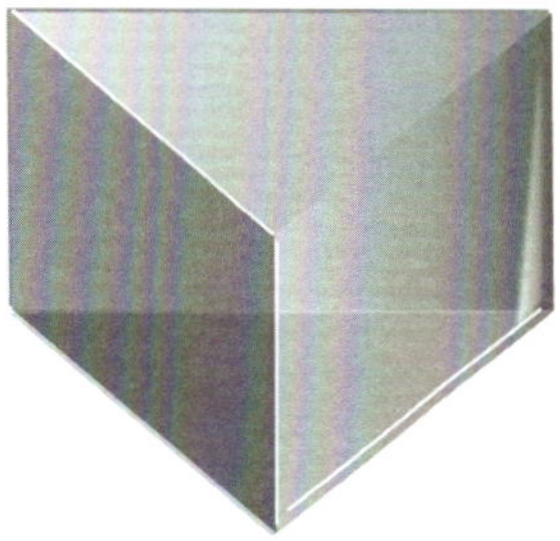

prism

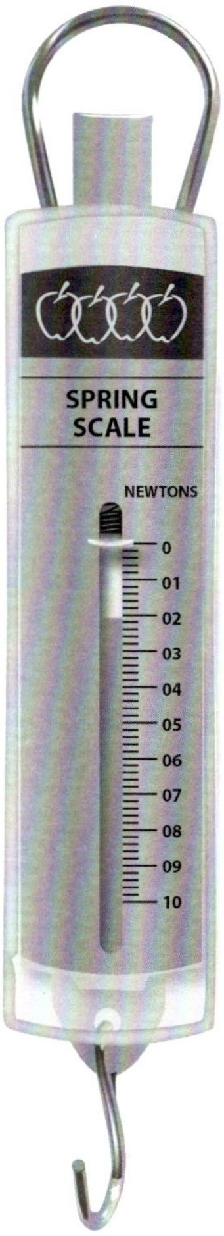

spring scale

timing device

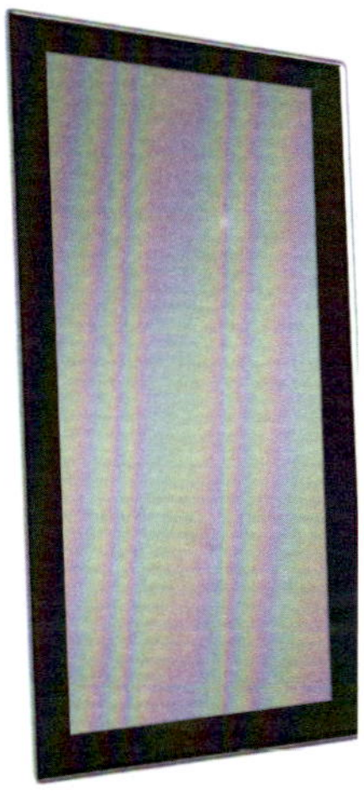

mirror

© REGION 4 EDUCATION SERVICE CENTER. ALL RIGHTS RESERVED.

UNIT 3: Force, Motion, and Energy

Lesson 1: Force and Motion

Let's Engage!

1. Take a silent walk around your school.

2. Record at least five actions that you observe.
3. Record at least five objects that are motionless.

Think about—

- What causes movement?
- Why are some objects motionless?

Let's Explore!

1. Rotate through the stations.
2. Read the station card before beginning the activity.
3. Follow the classroom safety rules.
4. Follow the instructions carefully.
5. Ask questions.
6. Make, share, and record observations.

Stations

1 Gravity

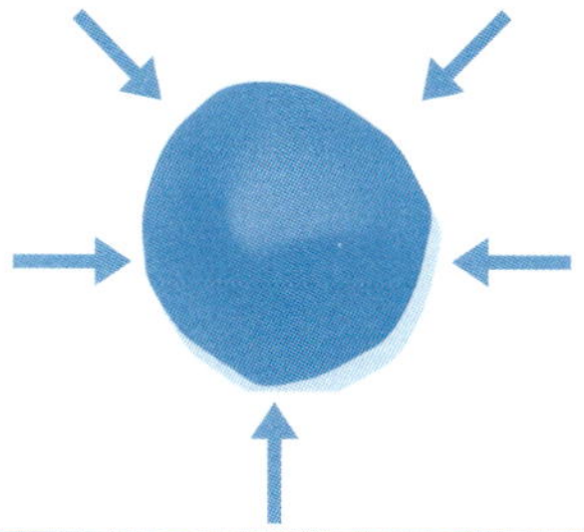

2 Magnetism

Let's Explain!

Force

You already know that a force is a push or a pull.

A force can make an object—

- start or stop moving
- move faster or slower
- change direction
- change shape

Try this:

1. Roll the clay into a ball.
2. Use a force to—
    - move the ball, then stop the ball
    - move the ball faster, then slower
    - change the direction of movement
    - change the shape of the ball

What type of force did you apply to the ball each time? A push or a pull?

© REGION 4 EDUCATION SERVICE CENTER. ALL RIGHTS RESERVED.

Using a Spring Scale

Force can be measured. The measure of the force of gravity on an object is called weight.

A spring scale is a tool used to measure force.

Force is measured in newtons (N).

Observe the picture of a spring scale.

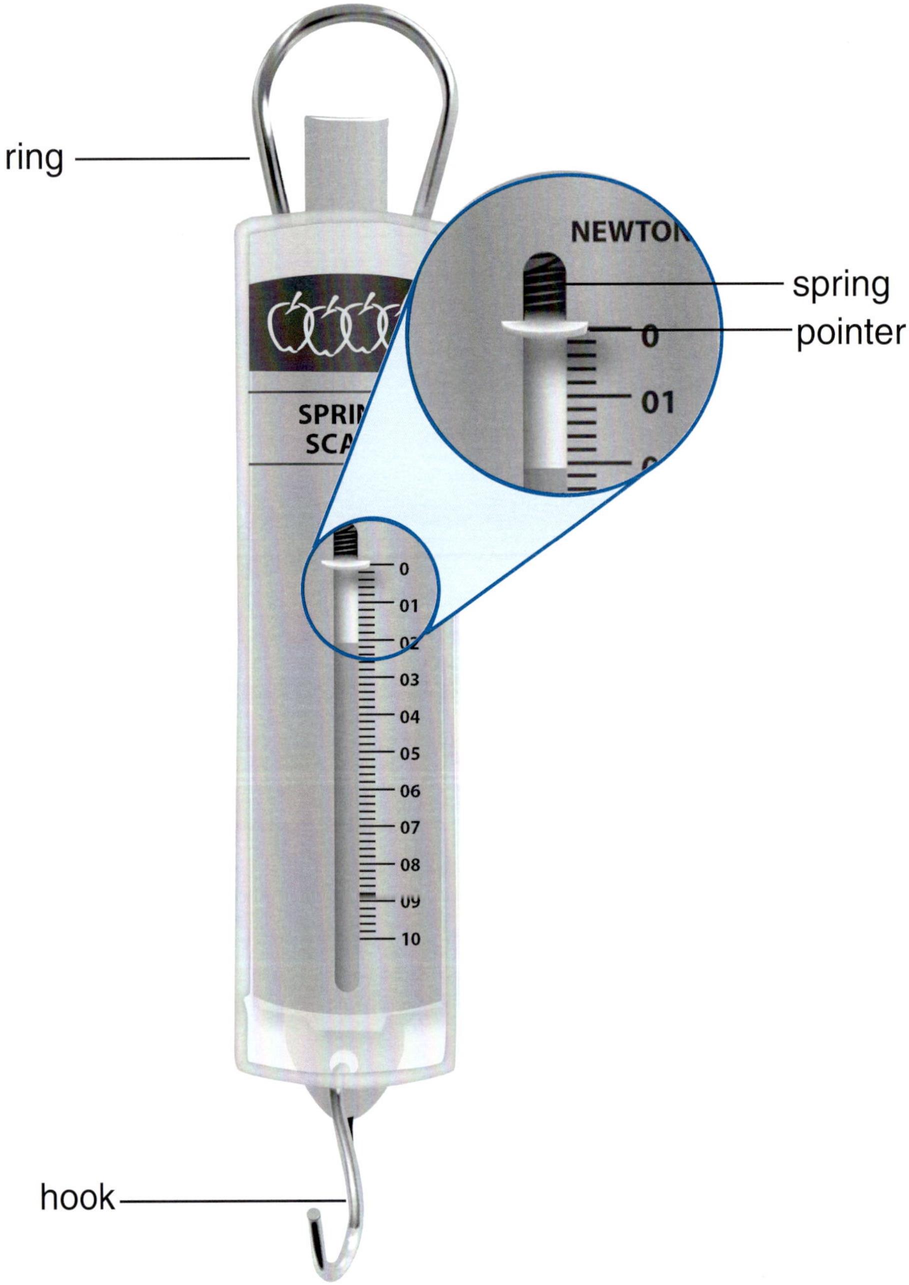

Observe the spring scale you will be using. How is it the same as the one in the picture? How is it different?

How do you measure the force of an object with a spring scale?

1. Hold the spring scale by the ring.
2. Check to make sure the scale reads 0 N (zero newtons).
3. Place the object on the hook.
4. Read the measurement at the top edge of the pointer.
5. Use newtons (N) as the unit of measurement.

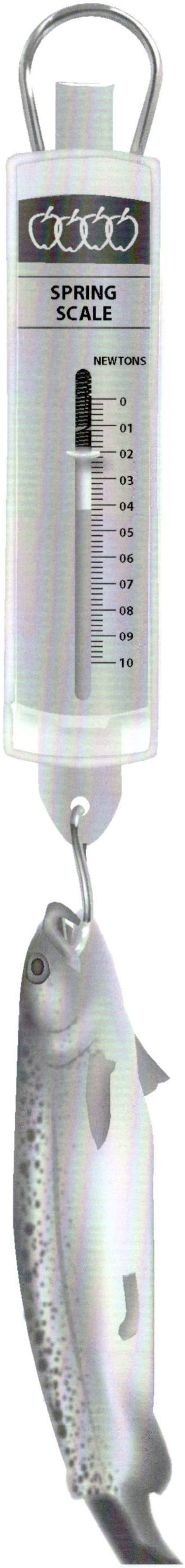

© REGION 4 EDUCATION SERVICE CENTER. ALL RIGHTS RESERVED.

The force of gravity on the bag of peanuts is 2 newtons (N).

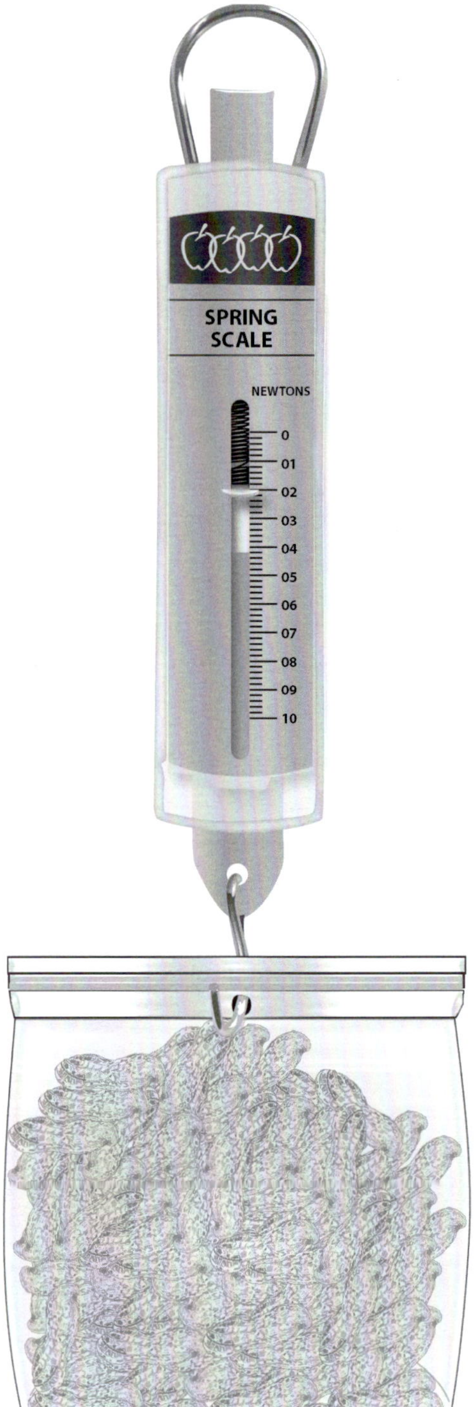

1. In your science notebook, draw the table below.
2. Use the spring scale to measure the pull of gravity on three objects.
3. Use the table to record your measurements.

Object Lifted	Amount of Force (N)

Friction

You already know that—

- Gravity is a force that pulls objects toward each other.
- Magnetism is a pulling force that attracts objects to a magnet.

Friction is also a force. It is produced when objects rub against each other. Mechanical energy is the energy an object has because of its motion or position. Friction is a force that works against motion. It can cause objects to slow down and stop moving.

The force of friction can be useful.

Have you ever tried to—

- Walk on an icy sidewalk? Sand may be spread on icy sidewalks to increase friction between your shoes and the icy surface. Friction helps to keep you from falling down.

© REGION 4 EDUCATION SERVICE CENTER. ALL RIGHTS RESERVED.

- Open a new jar of grape jelly? Using a rough kitchen towel to open the jar increases friction on the lid of the jar. The friction helps open the jar more easily.

- Throw a baseball? The surface of a baseball is rough. The friction between your hands and the surface of the baseball allows you to grip the baseball.

Friction also can produce heat energy.

Think about—

- A driver using brakes to stop a truck. Friction helps to stop the truck. Heat energy is produced between the tires and the road. The scorched surface of the tires may cause a bad odor.

Try this:

1. Place your hands on your face.
2. Make observations.
3. Rub your hands together for 10 seconds and place them on your face.
4. Make observations.

What happened to the temperature of the palms of your hands?

What caused the change in temperature?

Let's Elaborate!

Validating a Hypothesis

Scientists ask questions. They have ideas. Scientists try out, or test, their ideas.

Before testing their ideas, scientists make educated guesses about what might happen. These educated guesses are known as hypotheses.

A hypothesis is an idea based on existing information. A hypothesis may or may not be valid. An experiment can test whether a hypothesis is valid.

Discuss the following questions with your group.

1. What forces did you investigate in this lesson?
2. What materials did you use?
3. What effects do forces have on objects?
4. What force would you like to investigate?

© REGION 4 EDUCATION SERVICE CENTER. ALL RIGHTS RESERVED.

Create a hypothesis that tests the effects of force on an object.

For example—

All magnets attract objects from the same distance.

1. Design an experiment that tests the hypothesis.
2. Use the activity sheet to guide your group to complete the activity.
3. Be ready to share your experiment with the class.

Let's Evaluate!

Use your knowledge of force and motion to complete the assessment.

Let's Engage!

Follow your teacher's instructions.

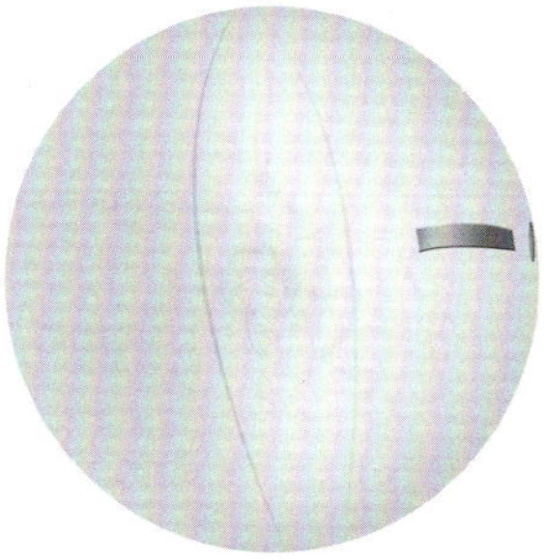

What caused the energy ball to produce light and sound?

What caused the energy ball to stop producing light and sound?

Let's Explore!

Activity 1

1. Read the procedure on the activity card.
2. Complete the task.
3. Be ready to share your observations and drawings.

© REGION 4 EDUCATION SERVICE CENTER. ALL RIGHTS RESERVED.

Activity 2

1. Read the procedure on the activity card.
2. Complete the task.
3. Be ready to share your observations and drawings.

Let's Explain!

Electrical Energy

Electrical energy is produced when charged particles move from one place to another.

A constant flow of charged particles is called an electrical current. A circuit is the path that an electrical current takes.

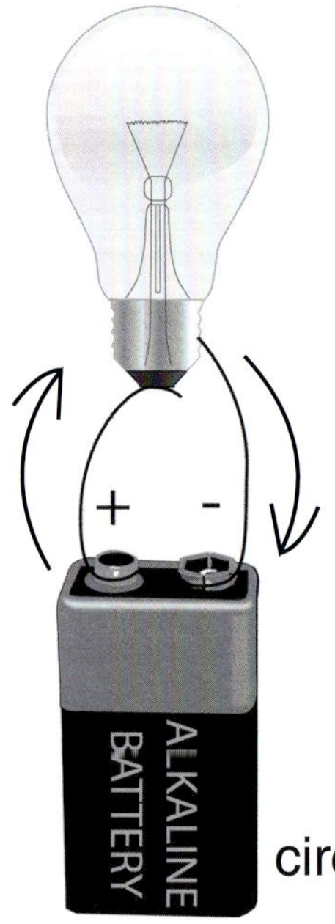

circuit

The source of the electrical energy for each circuit was a battery.

1. What items do you use that require a battery?
2. How do you use these items?

Let's Elaborate!

Stop Making Noise and Turn Off the Light

1. Read the procedure on the activity card.
2. Follow classroom safety rules.
3. Work together as a team to complete the activity.
4. Make, share, and record observations.

Electrical Power Plants

Think about—

- How does electrical energy get to your home?

Power plants produce electricity. They can produce electricity from other forms of energy, such as wind energy or nuclear energy. You will learn more about energy resources in upcoming lessons.

Electricity from a power plant is sent out, or transmitted, over power lines to homes and businesses. Electricity is then changed into other kinds of energy, such as light, heat, or mechanical energy.

© REGION 4 EDUCATION SERVICE CENTER. ALL RIGHTS RESERVED.

Create a list of five items you use at home or school that are plugged into an electrical outlet.

1. How do you use the item?
2. How does the electricity change?
3. During a storm, a neighborhood lost electricity. Use your knowledge of electrical energy to explain why this happened.

Let's Evaluate!

Use your knowledge of electrical energy to complete the assessment.

Let's Engage!

Make observations.

Think about—

- How does light travel?
- What happens to light when it interacts with different kinds of matter?

Let's Explore!

Activity 1

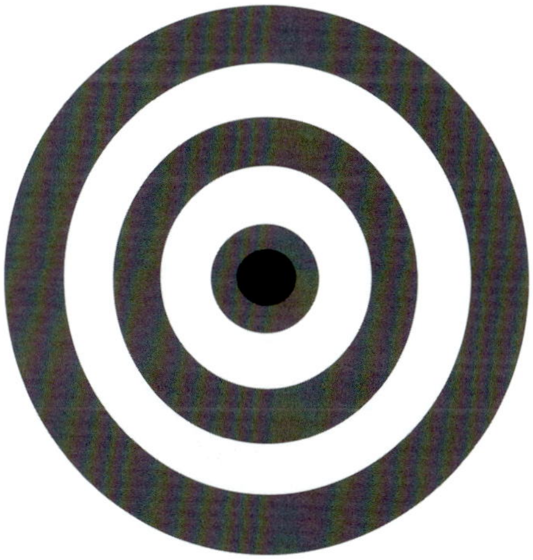

1. Read the procedure on the activity card.
2. Complete the task.
3. Be ready to share your observations and drawings.

© REGION 4 EDUCATION SERVICE CENTER. ALL RIGHTS RESERVED.

Activity 2

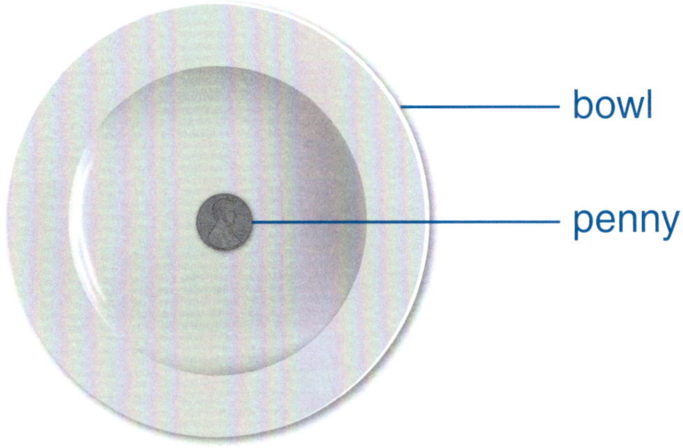

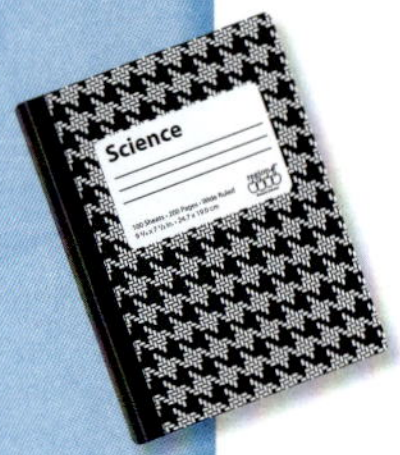

1. Read the procedure on the activity card.
2. Complete the task.
3. Be ready to share your observations and drawings.

Let's Explain!

Light Energy

The Sun is the major source of energy for Earth. Solar energy changes into different forms. One of those forms is heat energy. The other is light energy.

Light energy travels in straight lines called rays. Rays travel in straight lines until they strike an object.

What did you observe when a mirror was held in the beam of light? The light bounced off the mirror and traveled in a different direction. The bouncing of light off a surface is called reflection.

Reflected light energy allows us to use our sense of sight. For example, light energy reflects off a book you are reading. This reflected light enters our eyes, which allows us to see the words on the page.

How did you use reflection to hit the target?

Sometimes light passes through a substance. A medium is a substance through which light is transmitted.

For example, the beam of light was transmitted through the glass of water. The glass of water was the medium through which the light traveled.

What did you observe when the beam of light was transmitted through the glass of water? When light passes through a medium, the rays of light bend. The bending of light rays when transmitted through a medium is called refraction. The laser beam was refracted as it passed through the glass of water.

Try this:

1. Place the wooden ruler into the container of colored water.
2. Get eye level with the container.
3. Make and share observations.

Refraction causes the ruler to appear as though it is bent.

Refraction also allows you to see the penny in the bowl from a greater distance.

© REGION 4 EDUCATION SERVICE CENTER. ALL RIGHTS RESERVED.

UNIT 3: Force, Motion, and Energy

Lesson 3: Reflection and Refraction of Light

Copy the diagrams below.

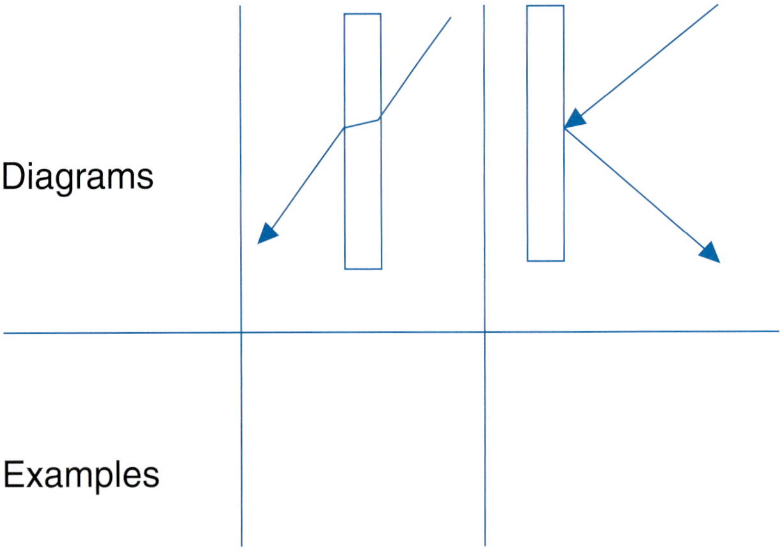

1. Which diagram represents reflection?
2. Which diagram represents refraction?
3. Label each diagram with the correct term and define each term in your own words.
4. List examples of each term under the corresponding drawing.

Let's Elaborate!

Reflection and Refraction of Light

Reflection and refraction of light can be observed in the natural world.

Photograph by Hanna Roach

The lighthouse and the trees along the shore of the pond are reflected on the water's surface.

When light is transmitted through raindrops, the light is refracted and separated into colors. If you are located where the light can reflect into your eyes, you will see a rainbow.

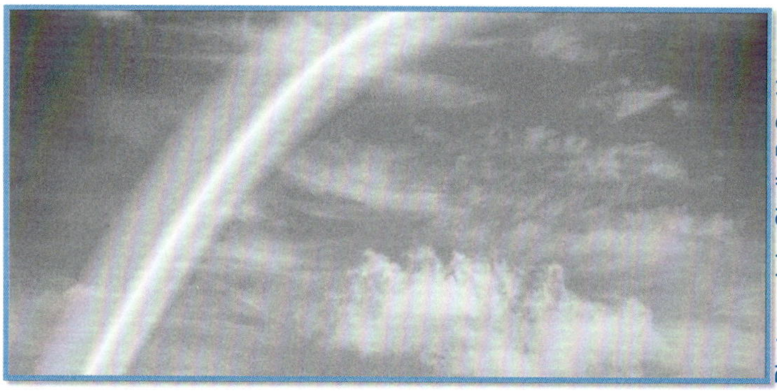

Photograph by Shelly D. Smith

1. Use a glass of water, a mirror, and a flashlight to model how refraction can form a rainbow.
2. Draw and label the parts of your refraction system.
3. Identify how light is reflected and/or refracted in the system.
4. Be ready to share your design with the class.

Reflection and refraction of light are an important part of things we use.

© REGION 4 EDUCATION SERVICE CENTER. ALL RIGHTS RESERVED.

UNIT 3: Force, Motion, and Energy

Lesson 3: Reflection and Refraction of Light

Choose one of the following items.

camera

binoculars

hand lens

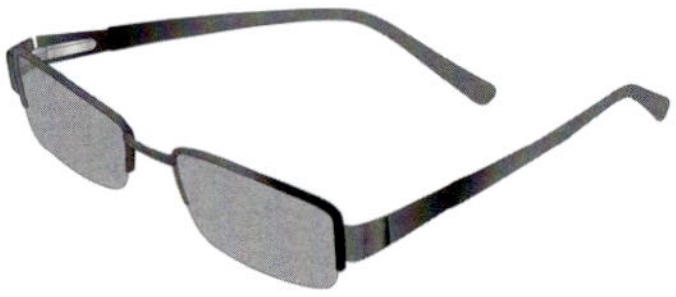

eyeglasses

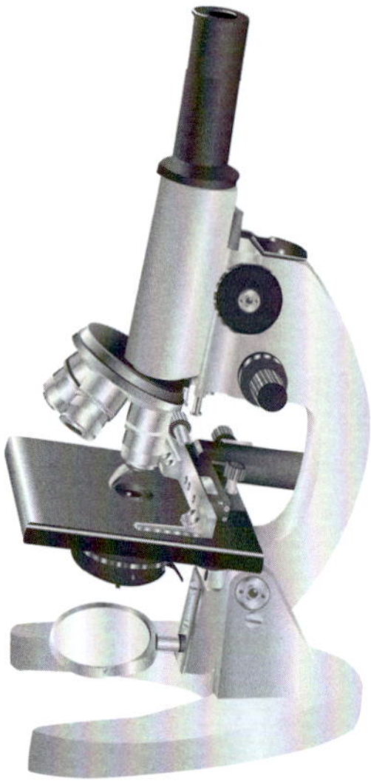

microscope

telescope

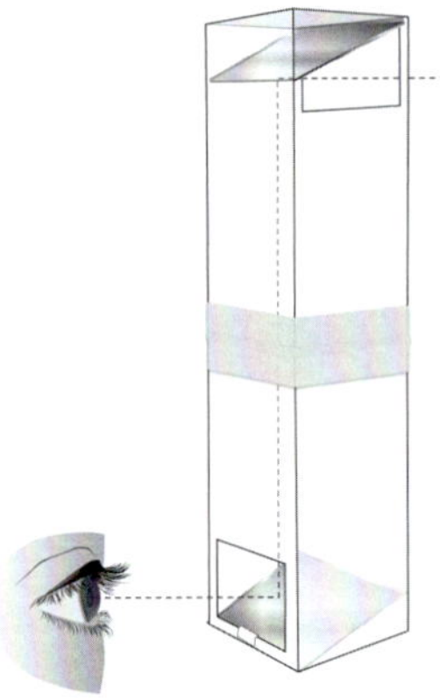

periscope

Use research tools, including the Internet and books in your school library, to determine how the object uses reflection or refraction.

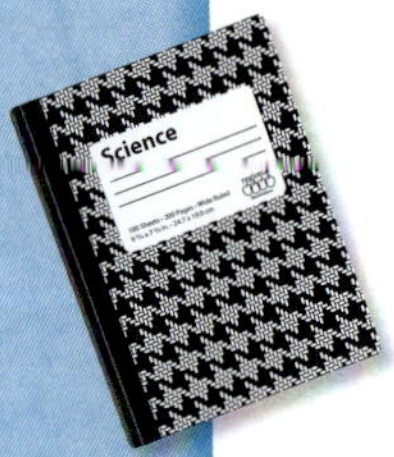

In your science notebook, draw and label a picture of the item.

Record how the object uses reflection or refraction.

Let's Evaluate!

Use your knowledge of reflection and refraction of light to complete the assessment.

Unit 4

Earth and Space, Part 1

UNIT 4: Earth and Space, Part 1

Introduction

Look at the night sky. What do you see? Stars? Planets? The largest star in our solar system is the Sun, the major source of energy for Earth.

By the end of this unit, you will—

Know these concepts:

- The Sun is a star that is the center of the solar system.
- Gravity is the force that holds the objects of the solar system in orbit.
- The Sun, Earth, and Moon form a system within the solar system.
- The Sun and oceans interact in the water cycle.
- Earth's rotation causes the day/night cycle.

Be able to do these things:

- Demonstrate safe practices in a science investigation.
- Make scientific observations.
- Make connections to prior knowledge.
- Communicate valid conclusions.
- Make models to represent the natural world.
- Identify the limitations of models.
- Describe interactions that take place in the Sun, Earth, and Moon system.
- Differentiate between weather and climate.
- Identify and compare the physical characteristics of the Sun, Earth, and Moon.
- Read tables.

Use these tools:

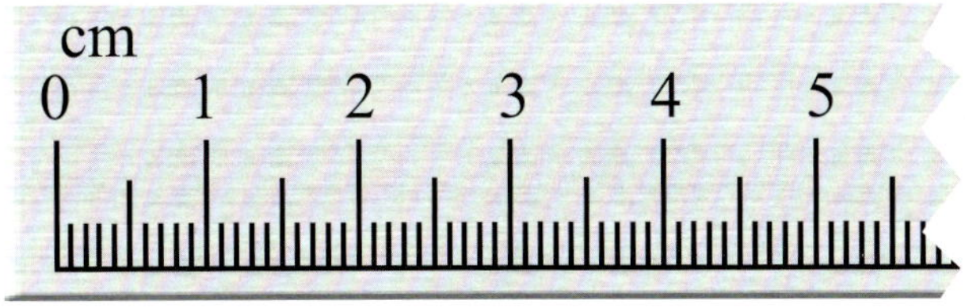

metric ruler

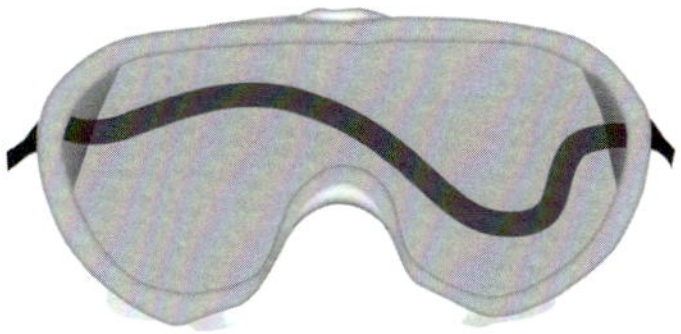

safety goggles

camera

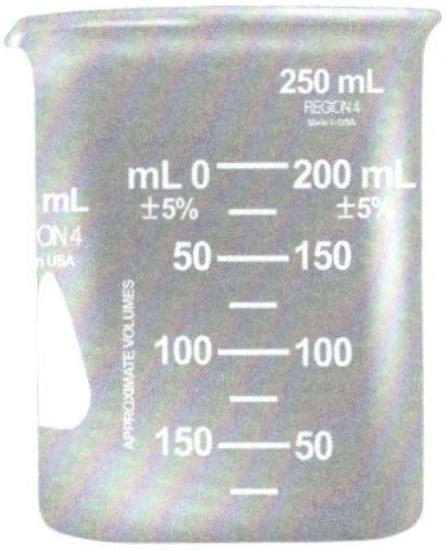

beaker

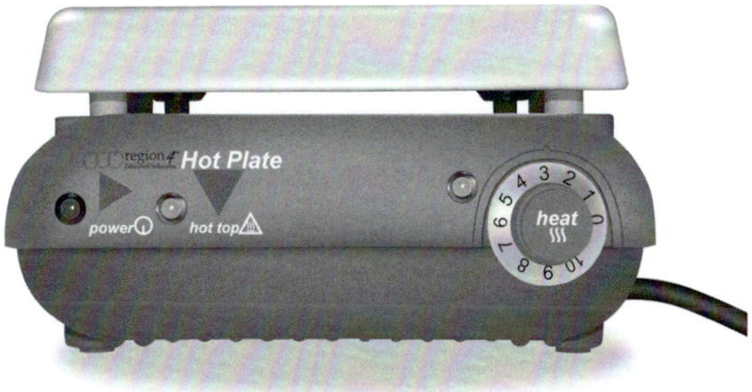

hot plate

© REGION 4 EDUCATION SERVICE CENTER. ALL RIGHTS RESERVED.

UNIT 4: Earth and Space, Part 1

Lesson 1: The Water Cycle

Let's Engage!

What do you already know about the water cycle?

Let's Explore!

Photograph by Shelly D. Smith

Read the procedure on the activity card.

Follow each step carefully to create a model of the water cycle.

Make, share, and record observations.

Let's Explain!

The Water Cycle

Water is constantly moving on Earth. A glass of water you drink today may be the water that fell as rain two weeks ago. The events that keep water moving create the water cycle.

In the water cycle, water moves from Earth's surface into the atmosphere. Water then moves back to Earth's surface. These events repeat over and over again.

One of the processes in the water cycle is precipitation. Precipitation happens when water falls from the clouds. It has many forms. Some forms of precipitation are rain, snow, sleet, and hail.

The water that falls from the atmosphere may soak into the surface of Earth and be stored as groundwater. It may also move across Earth's surface as runoff. As gravity pulls runoff down hills and slopes, the water collects in oceans and lakes. This is called collection.

Do you remember—

- Making a mixture of water and instant tea?
- Putting the mixture in a small container?
- Placing the container in a sunny location?
- Making observations of the mixture over time?

The process by which the mixture was separated is evaporation.

© REGION 4 EDUCATION SERVICE CENTER. ALL RIGHTS RESERVED.

UNIT 4: Earth and Space, Part 1

Lesson 1: The Water Cycle

Evaporation is a process in the water cycle. Evaporation occurs in the natural world when heat energy from the Sun turns water into a gas called water vapor. The water vapor then rises into the atmosphere. Plants also contribute to the water vapor in the air. A plant releases water vapor into the air through its leaves. This process is called transpiration.

Another process in the water cycle is condensation. Condensation occurs when the water vapor in the air cools. It then changes its state of matter from gas to liquid. The water droplets form clouds. The clouds become heavy. The water falls back to Earth as precipitation.

The cycle continues.

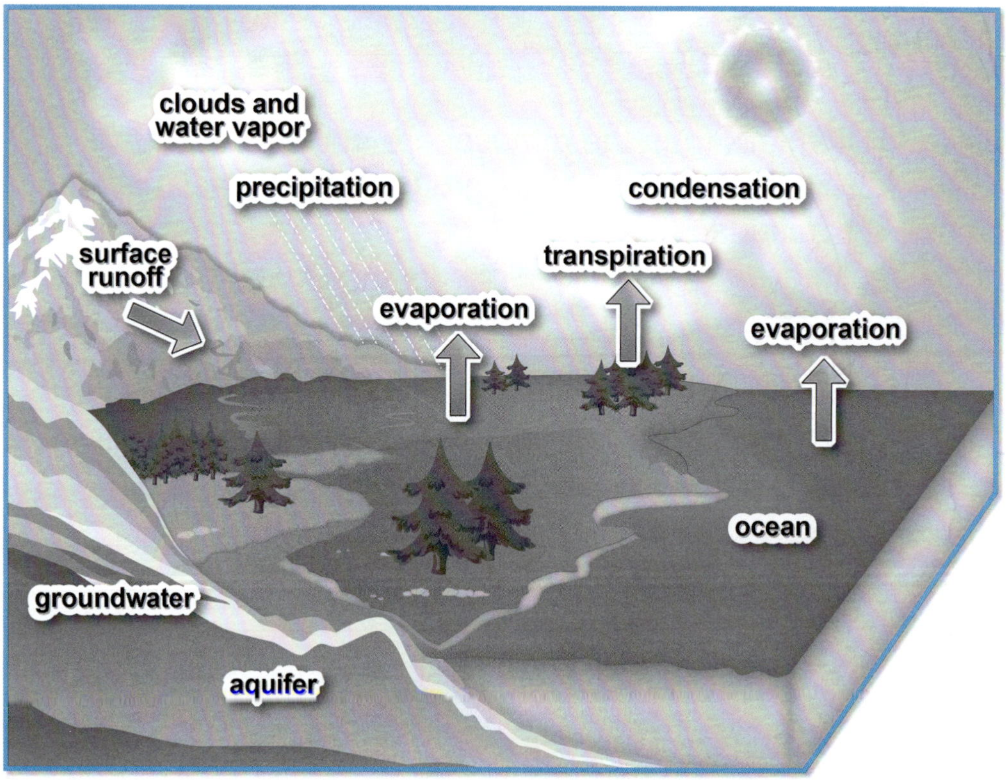

1. Review your diagram of the water cycle.

2. Revise the labeling of the diagram if needed.

3. Insert the diagram into your science notebook.

Photograph by Shelly D. Smith

The Role of the Sun and Oceans

Our Sun is the source of energy that drives the water cycle. Without a source of energy, there is no water cycle.

Earth's oceans hold 97% of the total water on the planet. The oceans are the source of 86% of the evaporation that occurs on Earth. As for precipitation, 78% of it occurs over oceans.

One of the basic needs of living things is water. The interaction of the Sun and the ocean in the water cycle is crucial for sustaining life on Earth.

© REGION 4 EDUCATION SERVICE CENTER. ALL RIGHTS RESERVED.

Let's Elaborate!

Weather and Climate

You have learned about precipitation. You have also learned about temperature. Precipitation and temperature determine the climate of an area.

Weather is the condition of the atmosphere at a certain time and place. Climate is the general weather of an area over a period of time.

Most scientists agree that there are three basic climates on Earth.

Use the information in Table 4.1 to answer the questions. Be ready to share your answers with the class.

Table 4.1. Earth's Climates

Climate	Temperature	Precipitation
polar	cold	dry
tropical	warm	wet or dry
temperate	changes from season to season	changes from season to season

Part 1

1. What type of temperatures does a polar climate have?
2. Does a polar climate have large amounts of precipitation?
3. What type of temperatures does a tropical climate have?
4. How is the precipitation in a tropical climate described?
5. What type of temperatures does a temperate climate have?
6. How is the precipitation of a temperate climate described?

Observe the location of each type of climate on Earth.

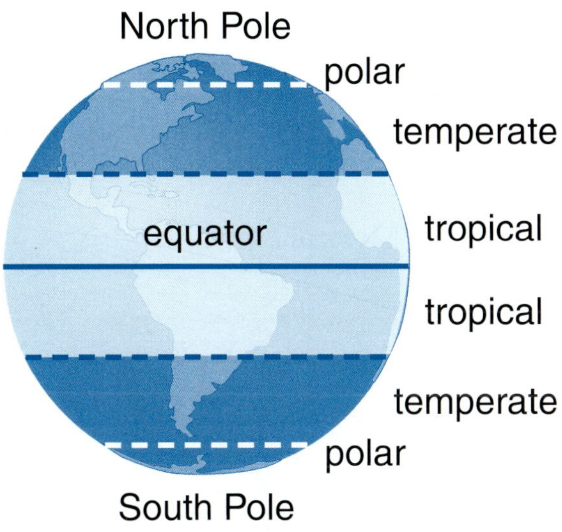

Part 2

1. In what climate do you live? Circle the label on the diagram your teacher gives you.
2. Insert the diagram into your science notebook.
3. Describe the climate where you live.
4. Does your description match the information in the table?

© REGION 4 EDUCATION SERVICE CENTER. ALL RIGHTS RESERVED.

Analyzing Data

Data is collected information. Scientists collect and study data. Data is often organized in graphs or tables.

Observe Graph 1.

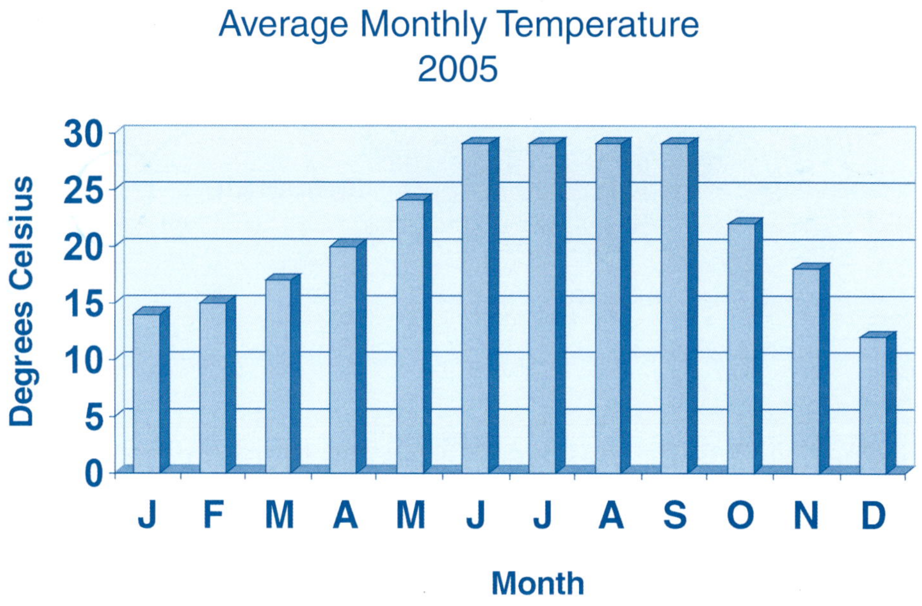

Use Graph 1 to answer the following questions.

1. What do the letters stand for at the bottom of the bar graph?
2. Was the average monthly temperature the same for each month? How can you know that by just making a quick glance at the graph?
3. In what months was the average monthly temperature the highest?
4. What was the temperature for those months?
5. In what climate was the data in Graph 1 most likely collected?

Observe Graph 2.

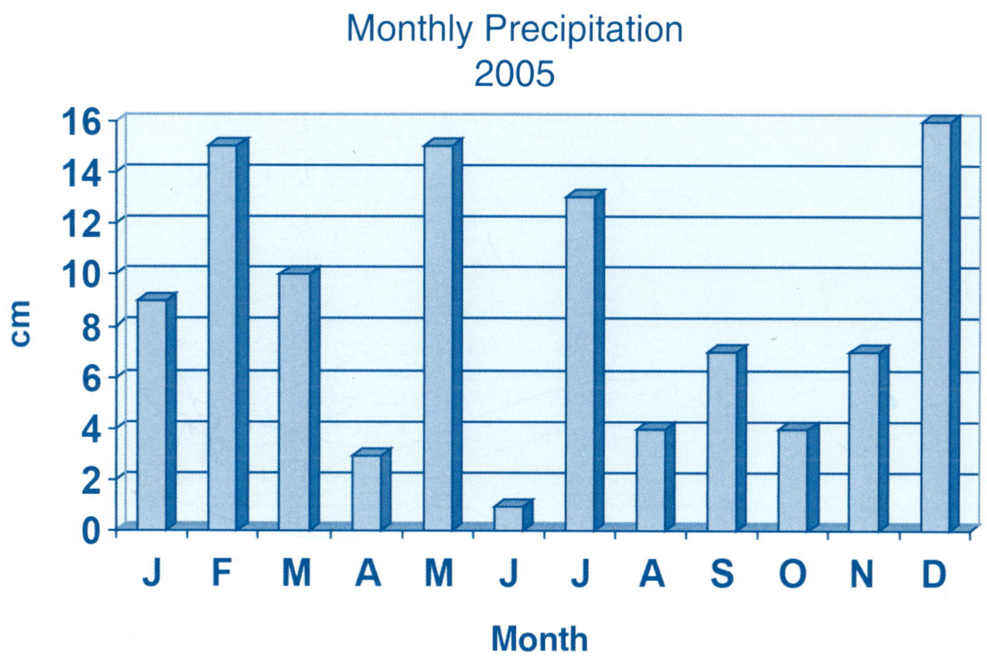

Use Graph 2 to answer the following questions.

1. Did the same amount of precipitation fall each month? How do you know?
2. Which month had the largest amount of precipitation?
3. How much precipitation fell that month?
4. Which month had the least precipitation?
5. How much precipitation fell that month?
6. In what climate was the data in Graph 2 most likely collected?

Let's Evaluate!

Use your knowledge of the water cycle, weather, and climate to complete the assessment.

© REGION 4 EDUCATION SERVICE CENTER. ALL RIGHTS RESERVED.

Let's Engage!

Let's Explore!

Making a Model

The diameter of Earth is four times larger than the diameter of the Moon.

The diameter of the Sun is 109 times larger than the diameter of Earth.

1. Use a meter stick and adding-machine tape to make a model comparing the diameters of the Sun, Earth, and Moon (scale: 1 cm = diameter of the Moon).
2. Record information in your folded booklet.
3. Display your model as instructed by your teacher.

Let's Explain!

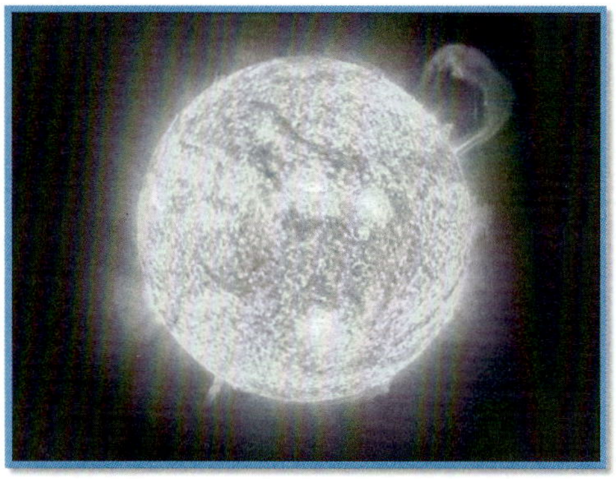

Photograph by NASA

The Sun

A star is an object in the sky made of very hot spinning gases. There are trillions of stars in space. The Sun is one of those stars.

The temperature of the Sun's surface is 5500 °C. It produces its own heat and light.

The Sun is located in the center of our solar system. In fact, it is the only star in our solar system. The Sun is classified as a medium-size star. It appears larger than other stars because it is the closest star to Earth.

The Sun's strong pull of gravity keeps objects in our solar system in orbit. An orbit is the path an object takes around another object. Earth revolves, or travels, in its orbit around the Sun once each year. Each trip around the Sun is called a revolution.

Record information about the Sun in your folded booklet after the class discussion.

Planet Earth

We live on the planet called Earth. It is the third planet from the Sun. Earth supplies the basic needs of living things—air, water, food, and shelter.

Earth is surrounded by a layer of gases called the atmosphere. Our atmosphere is a mixture of—

- 78% nitrogen;
- 21% oxygen; and
- 1% carbon dioxide, water vapor, and other gases.

© REGION 4 EDUCATION SERVICE CENTER. ALL RIGHTS RESERVED.

UNIT 4: Earth and Space, Part 1

Lesson 2: The Sun, Earth, and Moon

The atmosphere of Earth supports our need for oxygen in order to survive.

About 70% of Earth's surface is covered with water. Oceans hold about 97% of Earth's water in the form of salt water. Only about 3% of Earth's water is freshwater. Most freshwater is found in the form of glaciers, ice caps, and permanent snow. Usable forms of freshwater on Earth's surface are found in lakes, ponds, and rivers. You have already learned that the water on Earth is recycled through the processes of the water cycle.

About 30% of Earth's surface is land. The land is covered by a variety of landforms. A landform is a natural physical feature of Earth's surface. Mountains, craters, plateaus, plains, and valleys are examples of landforms found on Earth.

Record information about Earth in your folded booklet after the class discussion.

The Moon

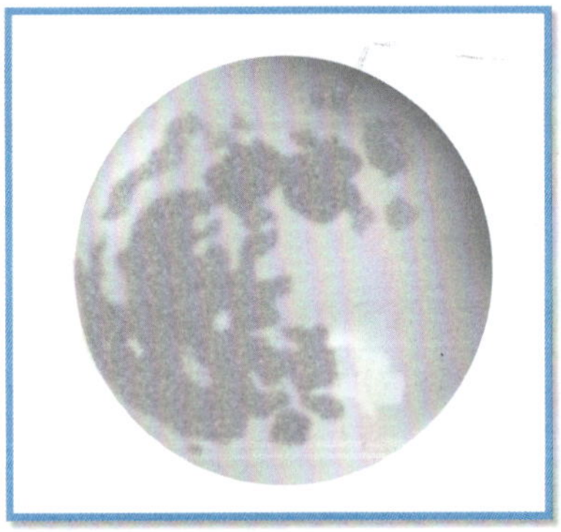

The Moon is Earth's only natural satellite. A satellite is an object that travels around another object. It takes the Moon 27.3 days to revolve around Earth. The Moon is about one-fourth the size of Earth, and it is about 348,000 kilometers away.

The Moon is made of rock. The rocky surface of the Moon has mountains, craters, flat areas, and valleys. The craters formed when meteorites struck the surface of the Moon. Meteorites do not burn away before reaching the Moon's surface because the Moon does not have an atmosphere.

Liquid water does not exist on the Moon, but there are small amounts of ice at the poles.

Living organisms do not exist on the Moon because it does not have an atmosphere to sustain life. The temperatures on the Moon can range from –160 °C to 100 °C.

You already know that—

The Moon does not produce its own light. It reflects the light of the Sun.

The Moon appears to change its shape from night to night. The part of the Moon that we see depends on the positions of Earth, the Moon, and the Sun at that particular time. These changes in the Moon's appearance are called the phases of the Moon, or the lunar cycle.

Neil Armstrong was the first man to walk on the Moon. Have you ever seen pictures of astronauts trying to walk on the Moon? The astronauts seem to bounce and float. What causes this?

Gravity is a force that pulls objects toward each other. The more mass an object has, the greater its force of gravity. The gravity on the Moon is much less than on Earth because it has less mass than Earth. Since the gravity on the Moon is less than the gravity on Earth, astronauts weigh less on the Moon. This is why they seem to float and bounce in space.

Record information about the Moon in your folded booklet after the class discussion.

© REGION 4 EDUCATION SERVICE CENTER. ALL RIGHTS RESERVED.

Let's Elaborate!

1. Discuss the following questions in groups:
    - How are the Sun, Earth, and Moon the same?
    - How are the Sun, Earth, and Moon different?
2. Create a poster of a triple Venn diagram to organize and display the information in your folded booklet.

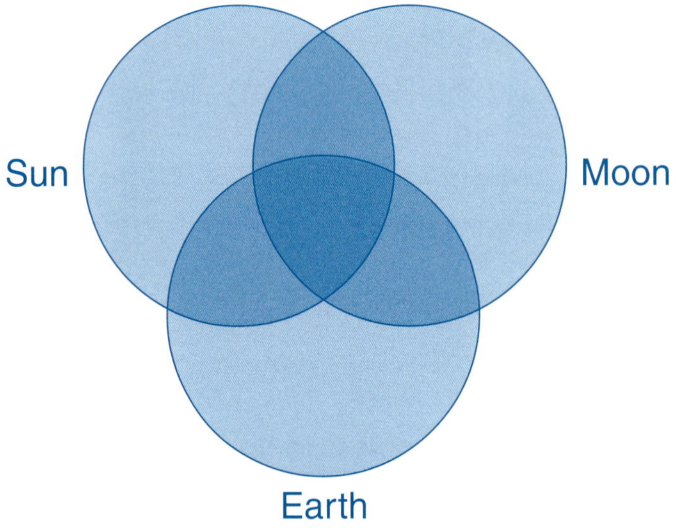

Let's Evaluate!

Use your knowledge of the Sun, Earth, and Moon to complete the assessment.

Let's Engage!

A student organizes her activities using a Venn diagram.

How are the activities classified?

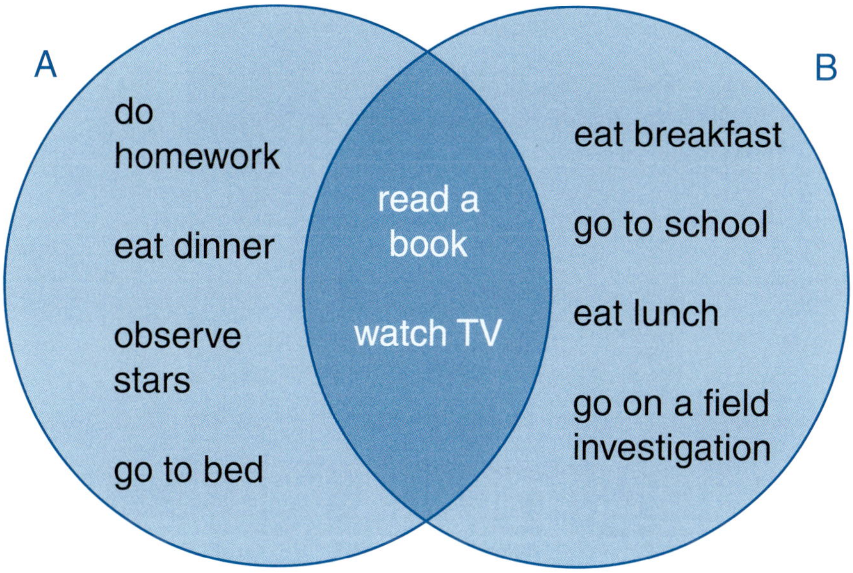

Let's Explore!

1. Choose a figure on the handout.
2. Cut out the figure.
3. Write your name on the figure.
4. Read all the steps on the activity card.
5. Complete the activity.

© REGION 4 EDUCATION SERVICE CENTER. ALL RIGHTS RESERVED.

Let's Explain!

The Rotation of Earth

Earth is continuously rotating, or spinning, on its axis. It takes Earth 24 hours to complete one rotation. The amount of time it takes Earth to complete one rotation is called a day. On Earth, one day equals 24 hours.

As Earth rotates on its axis, one half of Earth is having nighttime. The other half of Earth is experiencing daytime.

- At what point(s) is Florida having daytime?
- At what point(s) is Florida having nighttime?

Locate the state of Florida on the globe of Earth in the diagram.

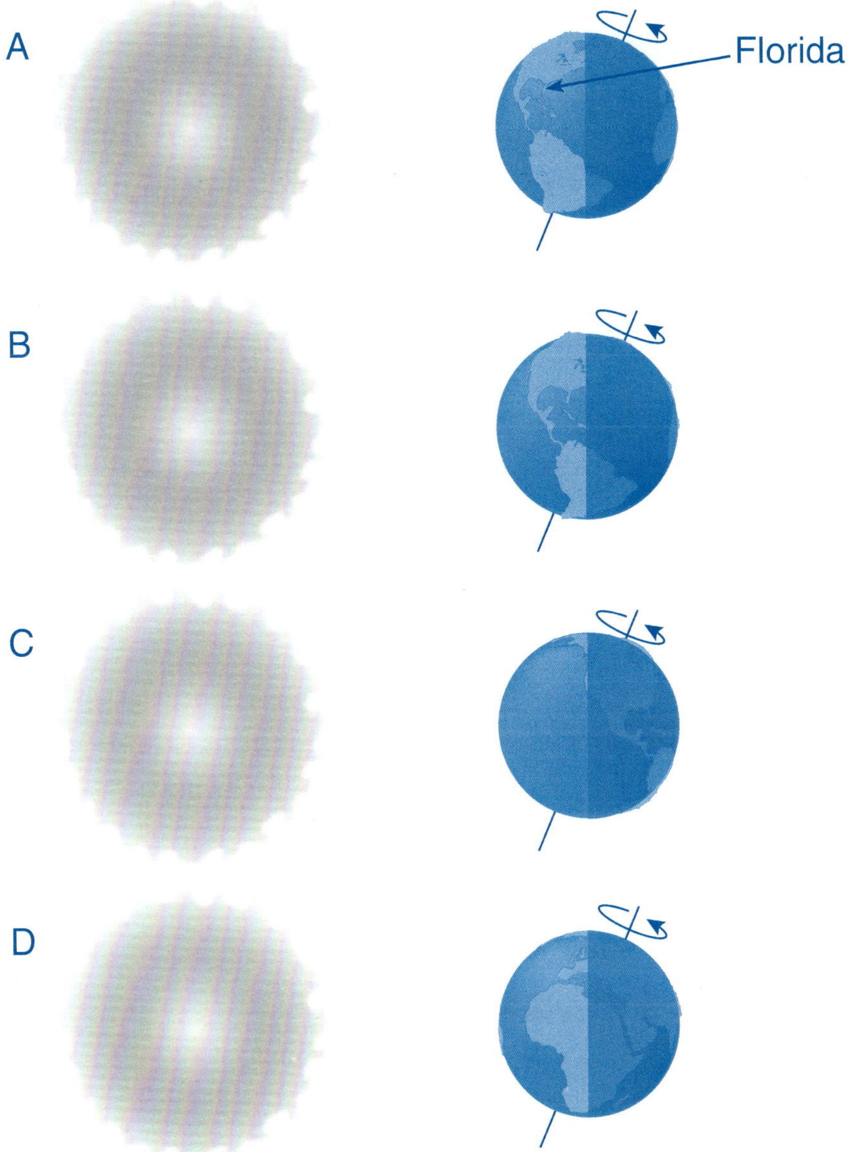

Sunrise and Sunset

Daytime begins at sunrise. Daytime ends at sunset.

From sunrise to sunset, the Sun appears to move across the sky. It is the rotation of Earth on its axis, however, that causes the Sun to appear as if it is moving.

© REGION 4 EDUCATION SERVICE CENTER. ALL RIGHTS RESERVED.

Observe the tables.

Table 4.2. Sunrise/Sunset Timetable June 2010 Dallas, Texas

Date	Sunrise	Sunset	Amount of Daytime
1	6:20 a.m.	8:30 p.m.	14 hours, 10 minutes
2	6:20 a.m.	8:31 p.m.	14 hours, 11 minutes
3	6:19 a.m.	8:31 p.m.	14 hours, 12 minutes
4	6:19 a.m.	8:32 p.m.	14 hours, 13 minutes
5	6:19 a.m.	8:32 p.m.	14 hours, 13 minutes
6	6:19 a.m.	8:33 p.m.	14 hours, 14 minutes
7	6:19 a.m.	8:33 p.m.	14 hours, 14 minutes

1. What is the title of the table?
2. What type of information is organized in the table?
3. What patterns do you observe in the table?
4. In what season of the year is the month of June?

Table 4.3. Sunrise/Sunset Timetable December 2010 Dallas, Texas

Date	Sunrise	Sunset	Amount of Daytime
1	7:12 a.m.	5:21 p.m.	10 hours, 9 minutes
2	7:12 a.m.	5:21 p.m.	10 hours, 9 minutes
3	7:13 a.m.	5:21 p.m.	10 hours, 8 minutes
4	7:14 a.m.	5:21 p.m.	10 hours, 7 minutes
5	7:15 a.m.	5:21 p.m.	10 hours, 6 minutes
6	7:16 a.m.	5:21 p.m.	10 hours, 5 minutes
7	7:16 a.m.	5:21 p.m.	10 hours, 5 minutes

1. What is the title of the table?
2. What type of information is organized in the table?
3. What patterns do you observe in the table?
4. In what season of the year is the month of December?

Compare the June 2010 table with the December 2010 table.

Make a valid conclusion comparing the amounts of daytime in June and December.

© REGION 4 EDUCATION SERVICE CENTER. ALL RIGHTS RESERVED.

Let's Elaborate!

1. Follow the safety rules for a field investigation.
2. Listen carefully to your teacher's instructions.

Evidence

Evidence is something that supports an idea.

For example—

Your grandmother asks you not to eat a peanut butter and jelly sandwich because it will spoil your appetite for dinner. You go ahead and make the sandwich, take it to your room, and eat it.

Later on, you go to the kitchen for dinner. Your grandmother asks, "Why did you eat that peanut butter and jelly sandwich?" She did not see you make the sandwich, take it to your room, or eat it.

How could your grandmother have known that you had eaten a peanut butter and jelly sandwich?

There was evidence.

Scientists collect evidence when they conduct investigations, make observations, and record data.

You investigated the day/night cycle and the movement of shadows. These two things are evidence that Earth rotates.

Let's Evaluate!

Use your knowledge of the rotation of Earth to complete the assessment.

© REGION 4 EDUCATION SERVICE CENTER. ALL RIGHTS RESERVED.

Unit 5

Earth and Space, Part 2

UNIT 5: Earth and Space, Part 2

Introduction

Have you ever seen the wind blow? Have you ever seen waves of water crash on a shoreline? Earth is constantly changing. Wind and water are just two of the forces that change the surface of Earth. Earthquakes occur. Volcanoes erupt. All these forces shape the surface of our planet.

Do you travel by car? Do you ride a school bus? Most types of transportation require fuel. Fossil fuels form beneath the surface of Earth. Fossil fuels are used to make gasoline, diesel fuel, and jet fuel.

By the end of this unit, you will—

Know these concepts:

- Forces of wind, water, and ice change the surface of Earth.
- New landforms can be a result of weathering, erosion, and deposition.
- Fossil fuels form over time from the remains of plant and animal life.
- Fossil fuels are found in sedimentary rock formations.
- Alternative energy resources are energy resources other than fossil fuels.
- Conservation of our natural resources is important.
- Fossils are evidence of past living organisms.

Be able to do these things:

- Demonstrate safe practices in a science investigation.
- Make scientific observations.
- Make connections to prior knowledge.
- Communicate valid conclusions.
- Make models to represent the natural world.
- Identify the limitations of models.
- Identify past events that led to the formation of Earth's natural resources.

Use these tools:

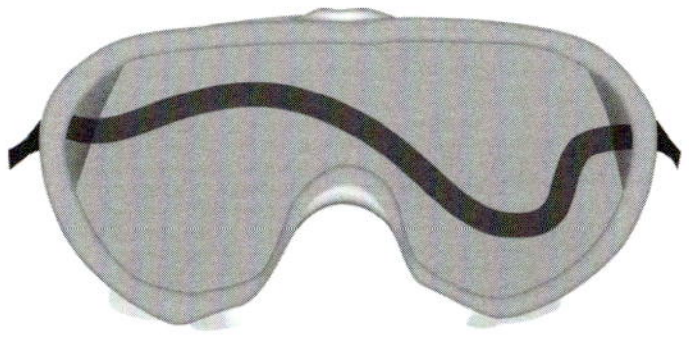

safety goggles

measuring spoons

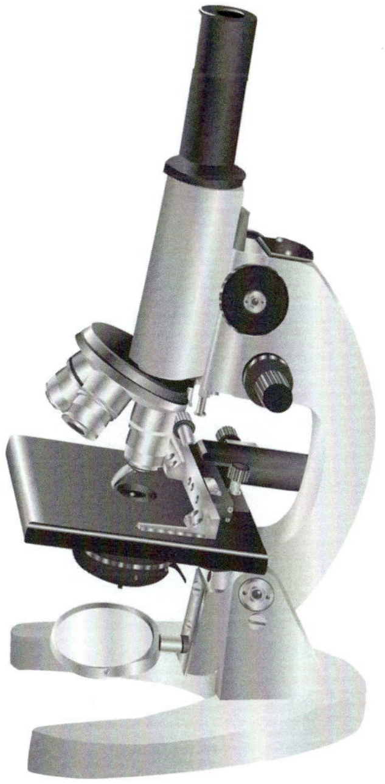

microscope

hand lens

© REGION 4 EDUCATION SERVICE CENTER. ALL RIGHTS RESERVED.

UNIT 5: Earth and Space, Part 2

Lesson 1: Earth's Changing Surface

Let's Engage!

Follow your teacher's instructions.

Photograph by Shelly D. Smith

Antelope Island, Utah

Let's Explore!

Stations

1 Force of Ice

2 Force of Wind and Water

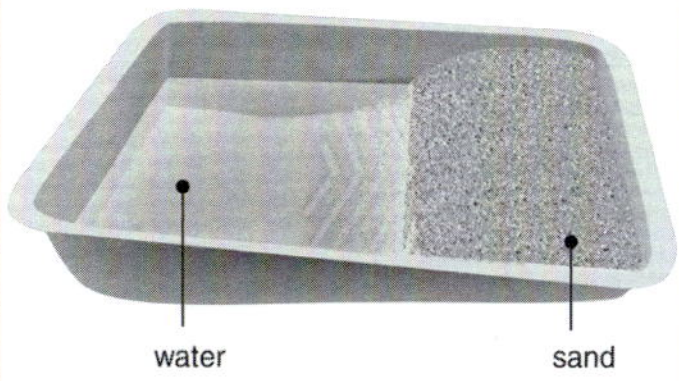

1. Follow the classroom safety rules.
2. Read the station card before beginning the investigation.
3. Follow the instructions carefully.
4. Make, share, and record observations.

Let's Explain!

Changing Earth's Surface

The forces of wind, moving water, and ice can break down, or weather, the surface of Earth. Sediment, or small pieces of rock, form as a result of weathering. Sediment is moved from place to place by wind, moving water, and gravity. This process is called erosion. Sediment is then dropped, or deposited, in a new location through the process of deposition. When the surface of Earth is changed by natural forces, new landforms may be created.

© REGION 4 EDUCATION SERVICE CENTER. ALL RIGHTS RESERVED.

The Force of a Glacier

A glacier is a large, slow-moving mass of ice. Glaciers form in very cold parts of the world. As a glacier moves, it slowly scrapes away rock sediment, uproots vegetation, and moves large boulders.

Over time, a glacier may create a new landform.

Observe the photographs.

Photograph by David Restivo, National Park Service

Glacial valley—Sun Point Overlook, Glacier National Park

- What shape is the glacial valley?

This mountain peak was carved away by glacial erosion.

Photograph by National Park Service

Glacial erosion—Thunderbird Mountain, Glacier National Park

The Force of Waves

A wave is a ridge of water that moves across a pond, lake, or ocean. Waves cause weathering and erosion along the shore. Waves can move large amounts of sand over time, building up sandbars and barrier islands.

A tsunami is a special kind of wave. It is a result of an earthquake. An earthquake releases a large amount of energy. The energy travels through the water and produces huge waves. The force of a tsunami can cause great damage and destruction to landforms. In fact, a tsunami can wash away entire islands.

In 1960, the fishing village of Queule, Chile, was hit by a tsunami.

Observe the photos of the city.

Photographs by NGDC, NOAA, US Department of Commerce

before

after

- What changes did the tsunami make to Earth's surface?

© REGION 4 EDUCATION SERVICE CENTER. ALL RIGHTS RESERVED.

Let's Elaborate!

Part 1

Photograph by Shelly D. Smith

Beach, Hawaii

1. Follow the classroom safety rules.
2. Read the activity card before beginning the activity.
3. Follow the instructions carefully.
4. Make, share, and record observations.

Sand Dunes

A sand dune is a mound of sand deposited by the wind. A sand dune can be observed along shorelines, in the desert, or at the base of a mountain. The formation of sand dunes requires three things:

- a source of sand
- wind
- a place for sand to be deposited

Observe the photograph.

Photograph by National Park Service

Great Sand Dunes National Park and Preserve, Colorado

The Great Sand Dunes were formed when large amounts of sand were carried by westerly winds and deposited at the base of the Sangre de Cristo range. The Great Sand Dunes are the tallest sand dunes in North America. The dunes' shape changes daily.

© REGION 4 EDUCATION SERVICE CENTER. ALL RIGHTS RESERVED.

UNIT 5: Earth and Space, Part 2

Lesson 1: Earth's Changing Surface

Sand dunes along a beach can change in a matter of hours.

There are two things that affect how fast a sand dune changes:

- wind speed
- size of the dune

Small dunes change more quickly than large ones. If vegetation is growing on the dune, weathering and erosion of the sand dune occur more slowly.

Part 2

1. Follow the classroom safety rules.
2. Read the activity card before beginning the activity.
3. Follow the instructions carefully.
4. Make, share, and record observations.

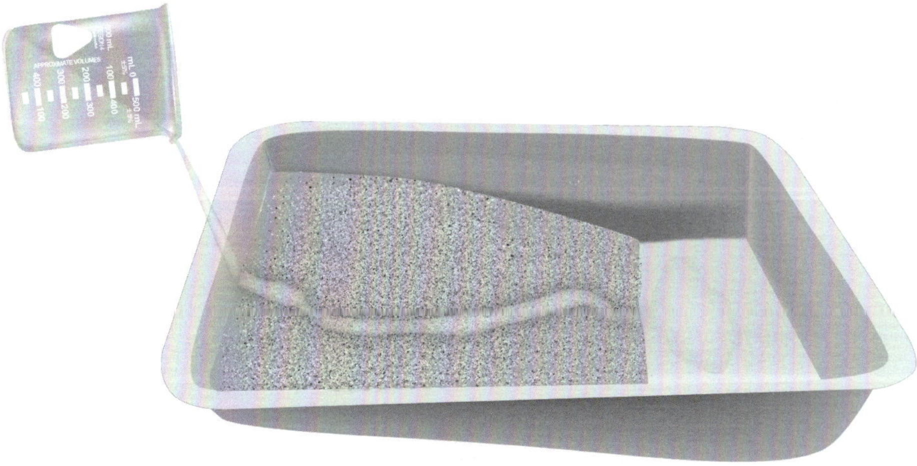

River Deltas

The moving water in a river causes weathering and erosion. Over very long periods of time, rivers can cut deep canyons into Earth's surface.

Observe the photograph.

Photograph by Jennifer Wilkening

Grand Canyon, Arizona

Many scientists agree that the Colorado River took 2 million years to carve the Grand Canyon in the Mojave Desert in Arizona.

Rivers can have other effects on Earth's surface. For example, a landform called a delta may form at the mouth of a river when the water slows down. A delta is a large, flat area of land that forms as sediment is deposited when a river enters a larger or slower moving body of water.

© REGION 4 EDUCATION SERVICE CENTER. ALL RIGHTS RESERVED.

Delta also is a letter of the Greek alphabet. The Greek symbol for delta is Δ.

River Delta

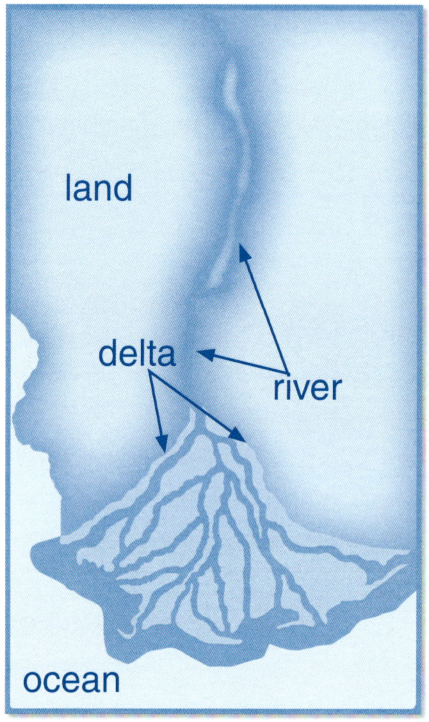

Observe the graphic.

Why do you think the landform is called a delta?

Earth's landforms are the result of a combination of destructive and constructive forces. Over time, parts of Earth's surface are torn down as other parts are built up.

- Choose one of the following natural forces:
    - water
    - wind
    - ice
- Describe how natural forces change the surface of Earth.
- Use the terms *weathering*, *erosion*, and *deposition* in the description.
- Identify landforms that are a result of natural forces.

Let's Evaluate!

Use your knowledge of Earth's changing surface to complete the assessment.

Let's Engage!

1. Listen closely to the story you are about to hear.
2. Close your eyes and picture the events in your mind.
3. In your science notebook, draw a picture of the events.

Let's Explore!

1. Follow the classroom safety rules.
2. Read the activity card before beginning the investigation.
3. Follow the procedure carefully.
4. Make accurate measurements.
5. Make, share, and record observations.
6. Think about—
   - What does each material represent in the natural world?
   - What does this activity model?
7. Draw and label the rock model.
8. Be sure to title your drawing.

© REGION 4 EDUCATION SERVICE CENTER. ALL RIGHTS RESERVED.

UNIT 5: Earth and Space, Part 2

Lesson 2: Sedimentary Rock

Let's Explain!

Types of Rock

There are three major types of rock.

- Igneous rock forms when melted rock cools and hardens.
- Sedimentary rock forms when layers of sediment are pressed and cemented together.
- Metamorphic rock forms when igneous and sedimentary rocks are heated and compressed over time.

1. Copy the table below into your science notebook.

Physical Properties of Rocks

Sample	Mass	Color	Texture	Does it have layers?	Does it contain fossils?
1					
2					
3					

2. Make and record observations of the rock samples.
   - Which sample(s) are sedimentary rocks?
   - What type of rock formation was modeled in the Explore activity?

Examples of sedimentary rocks are sandstone and limestone.

sandstone

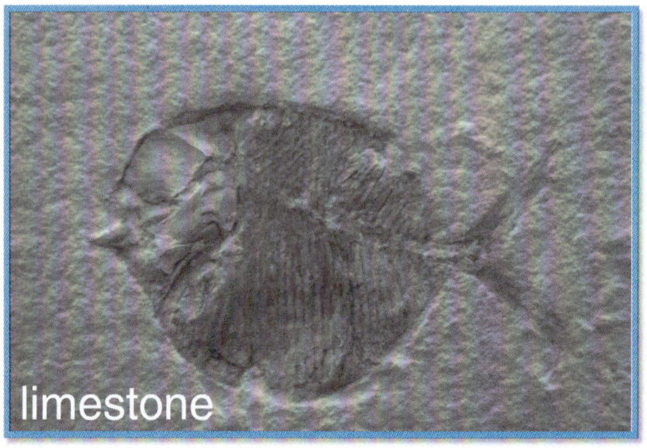
limestone

Photographs by Shelly D. Smith

Let's Elaborate!

Fossil Fuels

Fossil fuels are found in sedimentary rock formations.

Long ago, remains of prehistoric animals and plants were buried in layers of rock, mud, sand, or silt thousands of feet beneath Earth's surface. Over time, vast amounts of heat and pressure caused fossil fuels to form.

Petroleum, natural gas, and coal are fossil fuels. These fuels are sources of energy—energy that we use to heat our homes, cook our food, and travel in cars.

Petroleum is a liquid fossil fuel. It is found deep beneath Earth's surface. Natural gas collects underground above petroleum deposits. Wells are drilled into Earth's crust to obtain these fossil fuels.

© REGION 4 EDUCATION SERVICE CENTER. ALL RIGHTS RESERVED.

Coal forms in areas of heavy vegetation, such as swamps and forests. Coal is removed from Earth by the process of mining. Large holes are dug into the ground or underground to remove the coal from Earth's crust. Most of it is burned in power plants to produce electricity. Coal is used to produce about 50% of the world's electricity.

1. Read all the sentence strips.
2. Cut out the sentence strips.
3. Place them in the correct order to describe the formation of coal.
4. Insert the sentence strips into your science notebook.

Let's Evaluate!

Use your knowledge of sedimentary rock and fossil fuels to complete the assessment.

Let's Engage!

1. Use the materials in the resealable plastic bag to build a pyramid as shown below.
2. Using the materials in your bag, build as many pyramids as you can. Do not share materials with other groups.

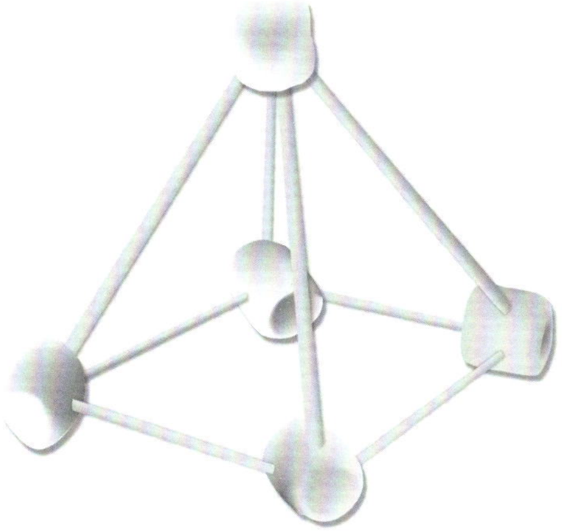

Let's Explore!

Make It Move

1. Find three ways to make the pinwheel turn.
   - Use only the materials provided by your teacher.
   - You may not touch the pinwheel blades with your hands or fingers.
2. Discuss with your group what the task and rules mean.
3. Develop a plan for completing the task.
4. Work together to make the pinwheel turn.

© REGION 4 EDUCATION SERVICE CENTER. ALL RIGHTS RESERVED.

Let's Explain!

Natural Resources

Natural resources are things found in the natural world that are useful to people. Some of those useful resources are found in Earth's crust.

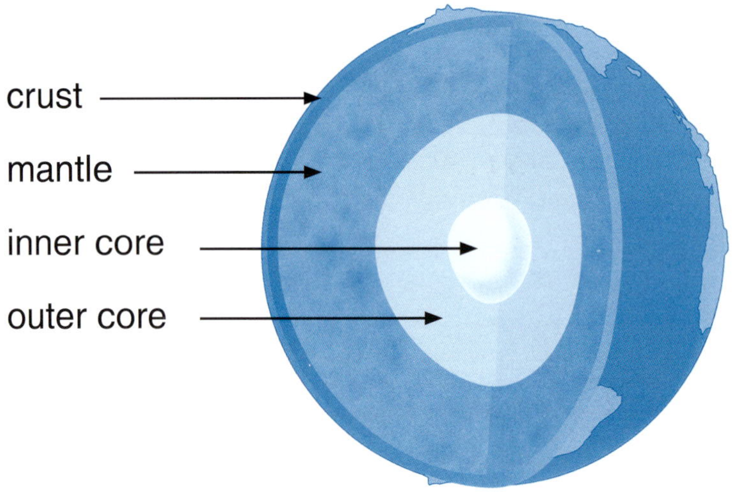

You have already learned that fossil fuels are—

- natural resources
- found in sedimentary rock formations deep inside Earth's crust
- major sources of energy

It may seem that our natural resources will never run out. There always seems to be electricity to use and fuel for cars. The fact, however, is that some natural resources are being used at a rate greater than the rate at which they can be renewed. This type of natural resource is called nonrenewable.

Since fossil fuels are formed over long periods of time and are not easily replaced, they are considered nonrenewable resources.

Alternative Energy Resources

The word *alternative* means a possibility, or substitute, for something else. Alternative energy resources are energy sources other than nonrenewable resources, or fossil fuels.

Wind

Wind is an alternative energy resource. An unlimited amount of wind energy exists on Earth.

Remember the pinwheel? Did you blow on the blades to make them turn? If you did, you used wind energy to make the pinwheel turn.

Today wind turbines are used to produce electricity. A group of wind turbines is called a wind farm.

wind turbines

How is electricity produced by a turbine? Wind causes the blades of a wind turbine to turn. The turning of the blades runs the generator, producing electricity.

© REGION 4 EDUCATION SERVICE CENTER. ALL RIGHTS RESERVED.

Solar Energy

You have already learned that the Sun is the major source of energy for Earth. The Sun provides enough energy in 1 minute to supply the world with energy for 1 year.

Energy from the Sun is called solar energy. Solar energy can be collected on solar panels. The collected solar energy is then changed into usable electrical energy.

solar panels

Geothermal Energy

Geothermal energy is heat energy found beneath Earth's crust. It collects in areas called reservoirs. Wells are drilled into geothermal reservoirs beneath Earth's surface. The hot water and steam from the reservoir go up through the wells to Earth's surface. The hot water and steam spin the turbines in a geothermal power plant. Electricity is produced.

Geothermal energy is an alternative energy resource.

Hydroelectric Energy

Hydroelectric energy is also an alternative energy resource. *Hydro-* is a prefix meaning water. Moving water contains energy. The amount of usable energy in moving water depends on how swiftly the water moves. Most often, a hydroelectric plant uses a dam to produce or generate electricity.

Think about—

- A river flows toward the ocean.
- A dam is built on the river.
- The dam holds back the water in the river.
- The water is collected in a reservoir.

Photograph by Karen Thompson

Lake Travis, Austin, Texas

Water is released from the reservoir. The moving water quickly flows through turbines at the base of the dam. The force of the moving water causes the turbines to turn.

Remember the pinwheel? Moving water caused the blades of the pinwheel to turn. As the turbines at the base of the dam turn, a generator converts, or changes, the energy of moving water into electricity.

© REGION 4 EDUCATION SERVICE CENTER. ALL RIGHTS RESERVED.

Biofuels

The prefix *bio-* means life. Biofuels are fuels made from plants.

Think about—

- Gasoline is a fuel that is burned in a car engine to make the car run.
- Have you ever seen the word *ethanol* on a gasoline pump? Gasoline may contain ethanol. Ethanol is a biofuel made from corn.
- Ethanol is an alternative energy resource.
- Another example of a biofuel is biodiesel. Biodiesel is a fuel that can be made from vegetable oils or fats. Biodiesel is an alternative energy resource.

Let's Elaborate!

Create a poster to persuade people to—

- conserve fossil fuels
- begin using alternative energy resources

Be prepared to present your poster to the class.

Let's Evaluate!

Use your knowledge of alternative energy resources to complete the assessment.

Let's Engage!

Use what you know about fossils to identify which of the following things are considered fossils.

- dinosaur tracks
- piece of ocean coral
- frozen woolly mammoth
- imprint of a leaf in stone
- piece of old pottery

Be prepared to justify your choices.

Let's Explore!

1. Follow classroom guidelines.
2. Read the procedures at each station before beginning the activity.
3. Work together to complete the activity.
4. Make, share, and record observations.

Stations

1 Formation of a Fossil

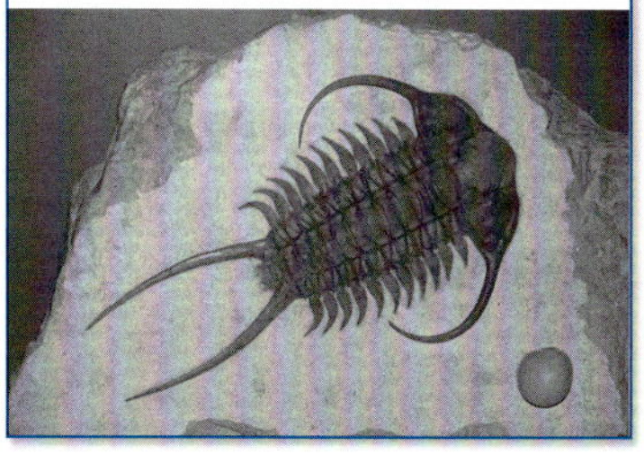

2 Formation of Petrified Wood

Photographs by Shelly D. Smith

ROTATION

© REGION 4 EDUCATION SERVICE CENTER. ALL RIGHTS RESERVED.

Let's Explain!

Evidence of Past Living Organisms

Evidence is something that shows proof of an idea or concept. Scientists interpret evidence to draw valid conclusions about an idea or concept.

A fossil is identified as evidence of a past living organism. It is evidence of what life was like before history was recorded in writing. The time before history was recorded in writing is called the prehistoric era. A scientist who studies the history of life on Earth is called a paleontologist.

leaf fossil

A fossil may be—

- The preserved remains of an animal. The frozen remains of a woolly mammoth are an example of this type of fossil. Remains of prehistoric insects can be preserved in ancient tree resin called amber.
- An impression of a prehistoric organism. An impression of a leaf in stone is an example.

- A cast of the mold left in stone when prehistoric organisms died and decomposed. This process was modeled at Station 1.
- The remains of a plant replaced by minerals. This process was modeled at Station 2.

Observe the diagram.

Final Event

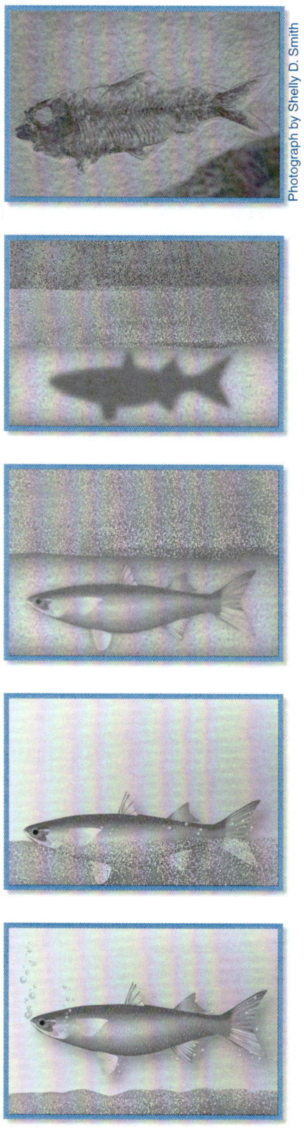

Scientists work at a dig site on land. They discover a fossil of a fish.

The remains of the fish rot, or decompose. Over time, layers of sediment are pressed together.

The fish is buried under many layers of sediment over a long period of time.

The fish dies.

A fish swims in the ocean.

Initial Event

The fossil of the fish is evidence that fish lived in prehistoric times and that the area once was covered with water.

© REGION 4 EDUCATION SERVICE CENTER. ALL RIGHTS RESERVED.

Fossils also give us clues about other animals that lived during prehistoric times. Evidence of animals called dinosaurs can be observed in fossils.

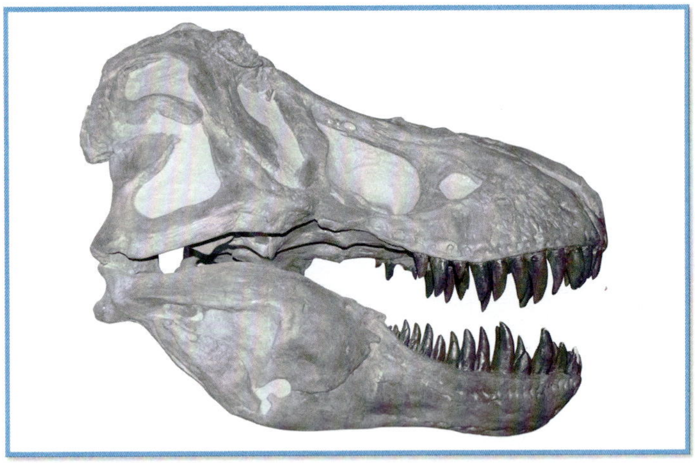

Photograph by Shelly D. Smith

This fossil shows that the dinosaur was most likely a meat eater. Its sharp teeth were used to capture and eat its prey.

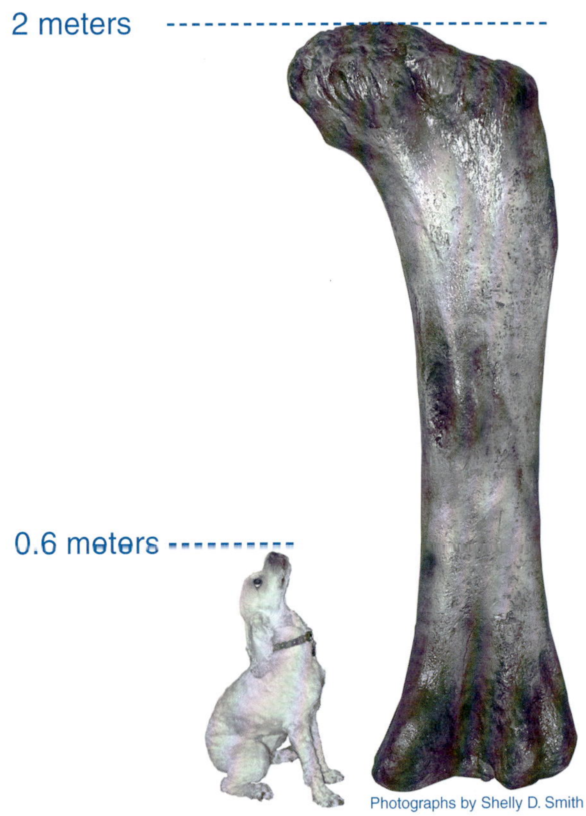

Photographs by Shelly D. Smith

This is a photograph of a fossil of a bone from the leg of a dinosaur. The fossil is a little more than 2 meters tall.

Try this:

Hold two meter sticks end to end vertically. This illustrates the height of the dinosaur's legs.

- What conclusions can you make about the size of the dinosaur?

Some animals living today look very much like animals that lived during prehistoric times.

The mammoth was a prehistoric animal. Many mammoths had long, curved tusks and long hair. Fossils of mammoths have been discovered in Europe, Asia, Africa, and North America.

- What animal living today looks like a mammoth?
- In what type of climate do you think the mammoth lived? Why?

Scientists are making discoveries of new evidence every day about Earth's history.

Research Project

1. Research a prehistoric animal of your choice.
    - In what type of environment did the animal live?
    - How did the environment meet the animal's needs?
2. Create a display showing the animal in its environment.

© REGION 4 EDUCATION SERVICE CENTER. ALL RIGHTS RESERVED.

Let's Elaborate!

Conservation of Natural Resources

To conserve means to protect or preserve. Conserving natural resources means to protect and use our natural resources wisely. You have already learned about the importance of conserving freshwater and using fossil fuels wisely.

Animals are natural resources. Animals are a source of food and clothing for people. If we are not careful of our interactions with animals, they may cease to exist and become extinct.

Why Are These Animals Extinct?

Read the passages below in your group.

Think about—

- What common reason do these animals share for becoming extinct?

The dodo was a bird. It had wings and feathers, but it could not fly. It lived on an island in the Indian Ocean. Dodos were easy to catch. They were a source of food for sailors and settlers who lived on the island. As the forests on the island were cleared, the habitat of the dodo was destroyed. The last dodo was seen in 1681.

dodo

The quagga was an animal related to the zebra. It lived in Africa. The quagga was hunted for its meat and its hide. The last quagga died in 1883 in a zoo.

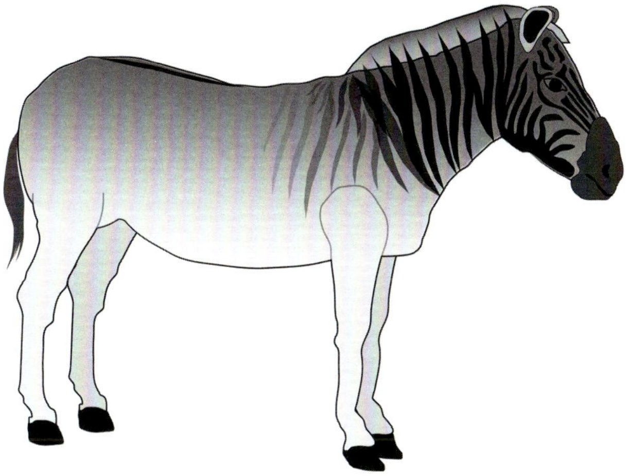

quagga

The Tasmanian wolf was not a wolf at all. It was closely related to the kangaroo. This animal lived in Tasmania, a state of Australia. In the 1800s, the Tasmanian wolf was thought to be a danger to livestock. Bounty hunters were paid to get rid of the animal. The Tasmanian wolf has been extinct since the 1940s.

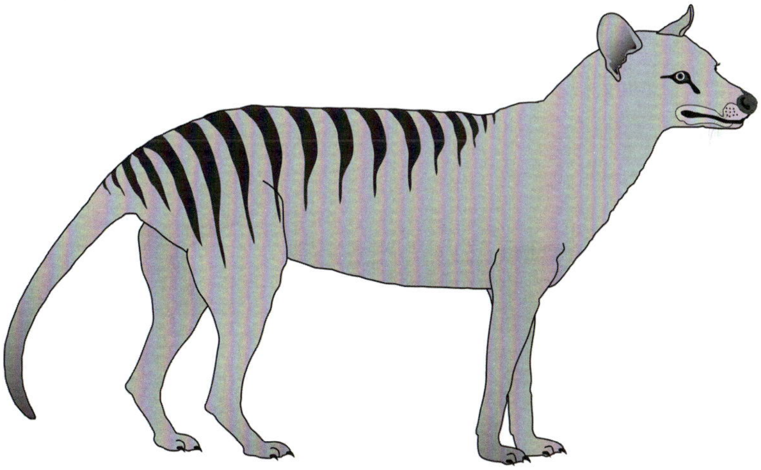

Tasmanian wolf

- What common reason do these animals share for becoming extinct?

© REGION 4 EDUCATION SERVICE CENTER. ALL RIGHTS RESERVED.

1. Why is it important to practice conservation of all our natural resources?

2. What can you do to preserve and protect our natural resources?

Let's Evaluate!

Use your knowledge of past living organisms and natural resources to complete the assessment.

Unit 6

Organisms and Environments, Part 1

UNIT 6: Organisms and Environments, Part 1

Introduction

Can a desert cactus survive and grow in a freshwater pond? Probably not. Living organisms survive and grow in an environment where they can meet their basic needs.

By the end of this unit, you will—

Know these concepts:

- Living organisms interact with living and nonliving parts of their environment.
- A food web shows the flow of energy between producers, consumers, and decomposers.
- The carbon dioxide-oxygen cycle is an exchange of gases between plants and animals.

Be able to do these things:

- Describe the flow of energy through a food web and food chain.
- Describe the process by which plants make their food.
- Identify the significance of the carbon dioxide-oxygen cycle.
- Predict the effects of changes in ecosystems by natural events and human interference.

Use these tools:

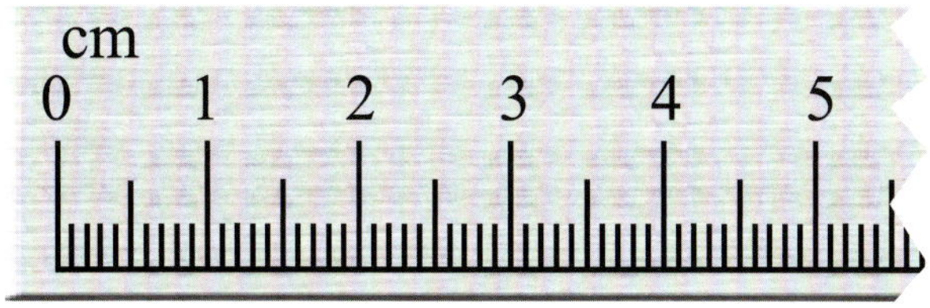

metric ruler

computer

timing device

terrarium

© REGION 4 EDUCATION SERVICE CENTER. ALL RIGHTS RESERVED.

UNIT 6: Organisms and Environments, Part 1

Lesson 1: Interactions of Organisms in Their Environments

Let's Engage!

Identify five natural resources.

- Which resources are living? Nonliving?

Let's Explore!

Research is learning and discovering facts about something. Library books, magazines, and the Internet are research tools.

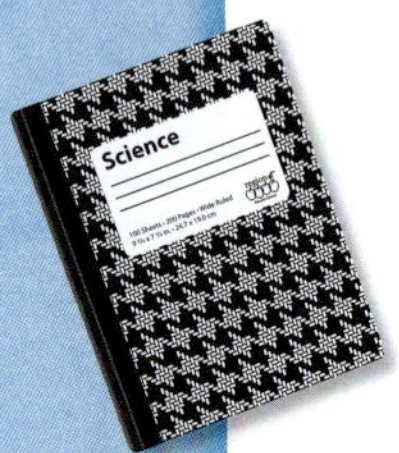

1. Copy Table 6.1. Animals in Nature into your science notebook.
2. Use research tools to learn facts about each animal and its environment.
3. Record facts to answer the question in each column.
4. Draw and label a picture of each animal.

Table 6.1. Animals in Nature

Animal	Ecosystem (Where is its habitat?)	Food (What does it eat?)	Protection (How does it protect itself?)
horned lizard			
bullfrog			
gray squirrel			
hammerhead shark			
American bison			

Let's Explain!

Ecosystems

The setting in which people, animals, or plants live is called an environment. An environment includes nonliving things such as air, water, sunlight, and soil. Living and nonliving things interact with each other in an environment to create an ecosystem.

There are many types of ecosystems. Forests, grasslands, deserts, ponds, marshes, and oceans are examples of ecosystems.

An ecosystem contains many different types of living organisms. Each organism needs enough food, water, and living space to survive. The place where an organism lives within an ecosystem is called its habitat.

Think about—

- A forest ecosystem may contain plants such as grasses, trees, and ferns. It may also contain animals such as beetles, earthworms, deer, squirrels, and rabbits. All the plants and animals in a forest ecosystem interact with each other.

For example—

- Grasses and trees provide food and shelter for birds and squirrels. The roots of plants hold soil and prevent it from eroding. Rabbits and deer graze on grasses and plants. Squirrels feed on acorns and spread seeds that produce new plants. Rabbits and squirrels are food for other animals.

Each part of an ecosystem helps keep the system in balance.

© REGION 4 EDUCATION SERVICE CENTER. ALL RIGHTS RESERVED.

UNIT 6: Organisms and Environments, Part 1

Lesson 1: Interactions of Organisms in Their Environments

Let's Elaborate!

Creating a Terrarium

1. Follow classroom and safety guidelines.
2. Read all the steps of the procedure before you begin.
3. Work as a group to complete the task.
4. Draw and label a picture of the terrarium.
5. In your own words, define and record the meaning of *terrarium*.
6. Create an organizer to record and classify the living and nonliving things in the terrarium.

Taking Care of Terrariums

Terrariums make good habitats for isopods and earthworms. Taking care of the terrarium is important for the survival of the living organisms.

- Keep the terrarium in an area that is well lit but does not receive direct sunlight. The terrarium needs to stay moist and cool at all times.
- Add more dry leaves as the animals eat them. Small pieces of potato and apple may also be added.
- Remove the lid and allow the terrarium to dry out slightly if mold begins to grow.
- Remove any rotting potato or apple.

Isopods

Isopods are not insects. They are related to animals such as crabs and lobsters. Isopods have seven plates that cover and protect their bodies. There are many types of isopods. One of the most common isopods is the pill bug, also known as the wood louse. Pill bugs protect themselves by rolling into a tight ball. This ball shape makes it harder for a predator to pick them up.

The pill bug breathes with gill-like structures. They must live in areas that are moist. Pill bugs are nocturnal. They prefer to hide during the day and come out at night when it is cool and dark.

Isopods have an important role in their ecosystem. They eat decaying plant material and are a food source for organisms such as ants, birds, and spiders.

Earthworms

Earthworms have segmented bodies that must stay moist. Earthworms are also nocturnal. They prefer to search for food at night when it is cool. During the day, earthworms burrow in the soil to find food where it is dark and moist.

Earthworms have an important role in their ecosystem. As the earthworm moves through the soil searching for food, it creates tunnels. These tunnels loosen the soil. This allows the roots of plants to easily push into the ground to absorb nutrients.

Earthworms also eat decaying plants. As they digest the plant material, nutrients from the plants are returned to the soil through earthworms' waste.

© REGION 4 EDUCATION SERVICE CENTER. ALL RIGHTS RESERVED.

- Discuss the following questions with your group.
- Record your answers in your science notebook.
- Be ready to share your answers during the classroom discussion.

  - How are isopods and earthworms similar?
  - How are isopods and earthworms important to their ecosystem?
  - How does the terrarium meet the specific needs of the isopod and the earthworm?

Let's Evaluate!

Use your knowledge of the interactions of organisms in their environments to complete the assessment.

Let's Engage!

What did you have for breakfast today?

Nutrition Facts

Serving Size		1 Cup (59g/2.1 oz.)
Servings Per Container		About 9
Amount Per Serving	**Cereal**	**Cereal with 1/2 Cup Vitamins A & D Fat Free Milk**
Calories	190	230
Calories from Fat	10	10
	% Daily Values	
Total Fat 1g	**2%**	**2%**
Saturated Fat 0g	**0%**	**0%**
Trans Fat 0g		
Polyunsaturated Fat 0.5g		
Monounsaturated Fat 0g		
Cholesterol 0mg	**0%**	**0%**
Sodium 250mg	**10%**	**13%**
Potassium 320mg	**9%**	**15%**
Total Carbohydrate 46g	**29%**	**29%**
Dietary Fiber 7g		
Sugars 17g		
Other Carbohydrate 22g		
Protein 5g		

INGREDIENTS: WHOLE-GRAIN WHEAT, RAISINS, WHEAT BRAN, SUGAR, HIGH-FRUCTOSE CORN SYRUP; CONTAINS 2 PERCENT OR LESS OF SALT, MALT FLAVORING, INVERT SUGAR, CORN OIL, NIACINAMIDE, REDUCED IRON, PYRIDOXINE HYDROCHLORIDE (VITAMIN B_6), ZINC OXIDE, RIBOFLAVIN (VITAMIN B_2), THIAMIN HYDROCHLORIDE (VITAMIN B_1), VITAMIN A PALMITATE, FOLIC ACID, VITAMIN D, VITAMIN B_{12}.

CONTAINS WHEAT INGREDIENTS.

© REGION 4 EDUCATION SERVICE CENTER. ALL RIGHTS RESERVED.

Let's Explore!

1. Choose two ecosystems.
2. Use your research to create a food chain for each ecosystem.
3. Think about—
   - What information do you need to create a food chain?
   - If you do not have all the information you need, where will you find the information?

Let's Explore!

Let's Explain!

Food Chains: Producers

A food chain shows the flow of energy from one organism to another. Both plants and animals need energy to grow.

Plants are called producers because they use energy from the Sun to produce, or make, their own food.

Observe the diagram.

Photosynthesis

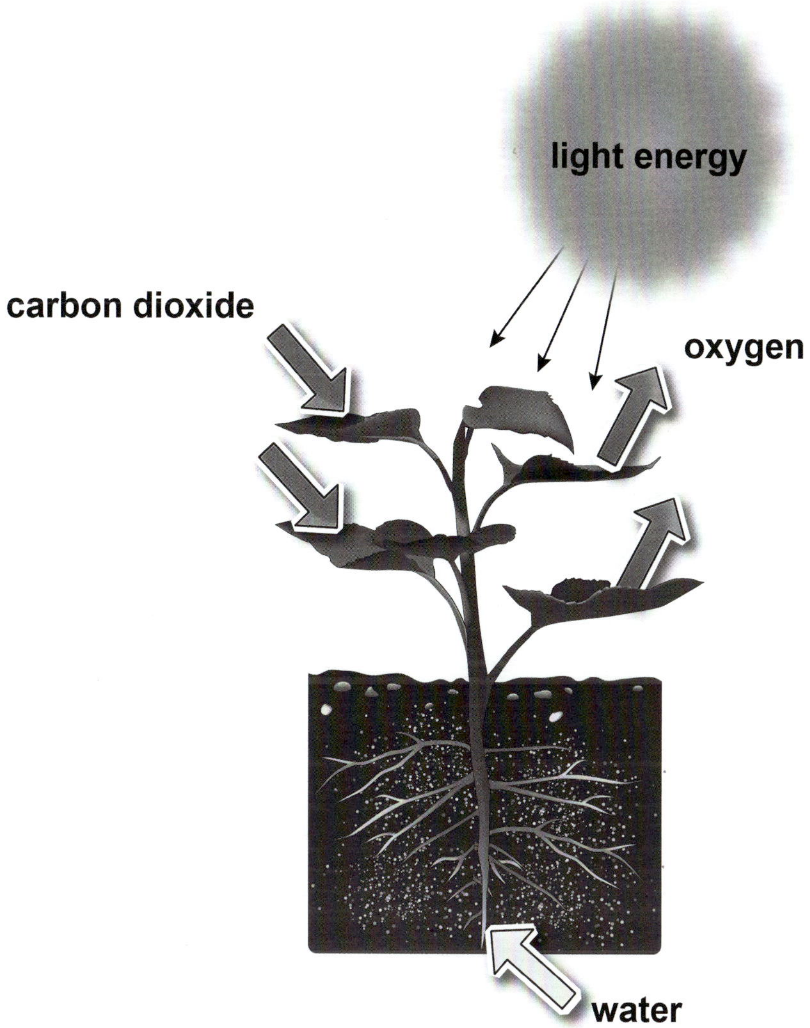

Light energy from the Sun is captured in the leaves of a plant. Water is absorbed by its roots and travels to the plant's leaves. The leaves also take in carbon dioxide from the air. With these three things—sunlight, water, and carbon dioxide—the plant makes its own food. This food is called glucose.

Sometimes plants make more food than they need. The extra food is stored in the roots and stem of the plant. The oxygen that is not used by the plant is released into the air. The process of using the Sun's energy to make food is called photosynthesis.

© REGION 4 EDUCATION SERVICE CENTER. ALL RIGHTS RESERVED.

UNIT 6: Organisms and Environments, Part 1
Lesson 2: Food Chains and Food Webs

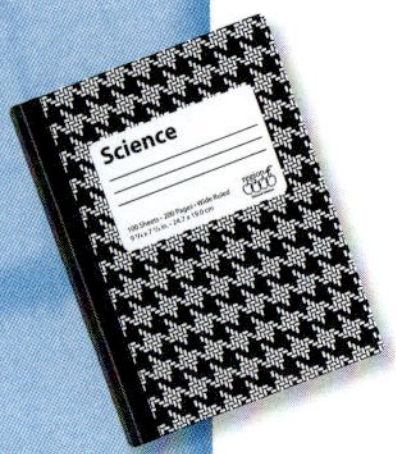

The first organism in a food chain is always a producer.

Draw and label the photosynthesis diagram.

Food Chains: Consumers

Animals get energy by eating plants or other animals. Since animals consume, or eat, other organisms, they are called consumers.

There are three types of consumers: herbivores, carnivores, and omnivores.

Herbivores

Herbivores eat only plants or plant products. They may eat fruits, flowers, or roots of plants. Cows, squirrels, and grasshoppers are herbivores.

grasshopper

Carnivores

Carnivores are animals that eat only other animals. Carnivore means "meat eater." Hawks, lions, eagles, and wolves are carnivores. Even a ladybug is a carnivore. Ladybugs eat aphids, or small bugs that eat plants. An animal that hunts and eats other animals is a predator. The animal that is hunted and eaten is the prey.

eagle

Omnivores

Omnivores eat both plants and animals. A grizzly bear is an omnivore. It will eat fish, berries, and corn. Crows, raccoons, and most humans also are omnivores.

raccoon

© REGION 4 EDUCATION SERVICE CENTER. ALL RIGHTS RESERVED.

Observe the food chain below.

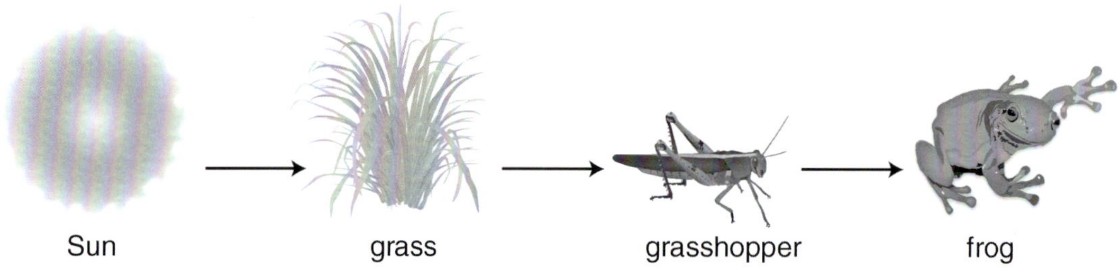

The Sun provides energy for the grass to grow. The grasshopper eats grass. The grass provides energy for the grasshopper. The frog eats the grasshopper. The grasshopper provides energy for the frog.

- What part(s) of the food chain are producers?
- What part(s) of the food chain are consumers?
- What types of consumers are in the food chain?

Food Chains: Decomposers

When plants and animals die, organisms called decomposers break down the remains of plants and animals into smaller pieces. Earthworms and bacteria are decomposers.

Roots of plants absorb nutrients from the decaying organisms. Plants use the nutrients to grow. The food chain continues.

Let's Elaborate!

Food Webs

You have already learned that a food chain shows the flow of energy from one living organism to another.

Observe the food chains below.

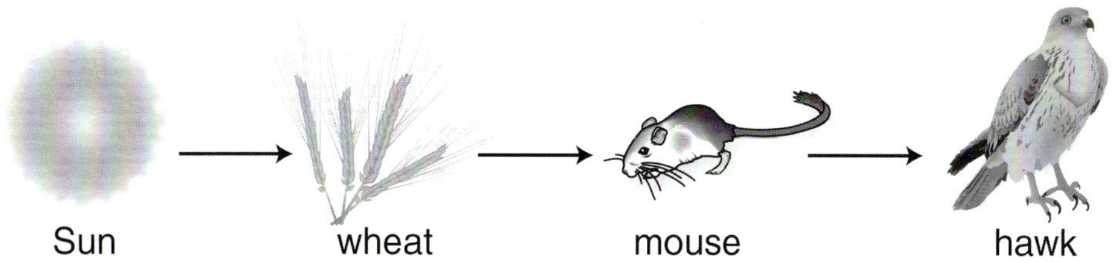

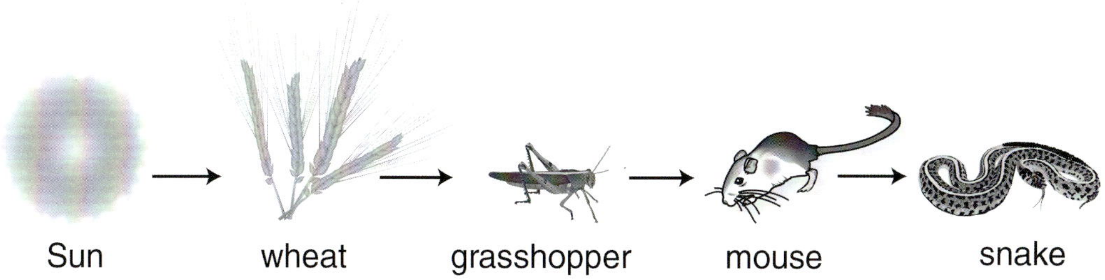

- What do these two food chains have in common?
- What type of consumer is a mouse?
- What evidence is shown in the food chains?

There may be many food chains in an ecosystem. The flow of energy can overlap, creating a food web.

© REGION 4 EDUCATION SERVICE CENTER. ALL RIGHTS RESERVED.

Observe the simple food web below. It was created from the food chains shown on the previous page.

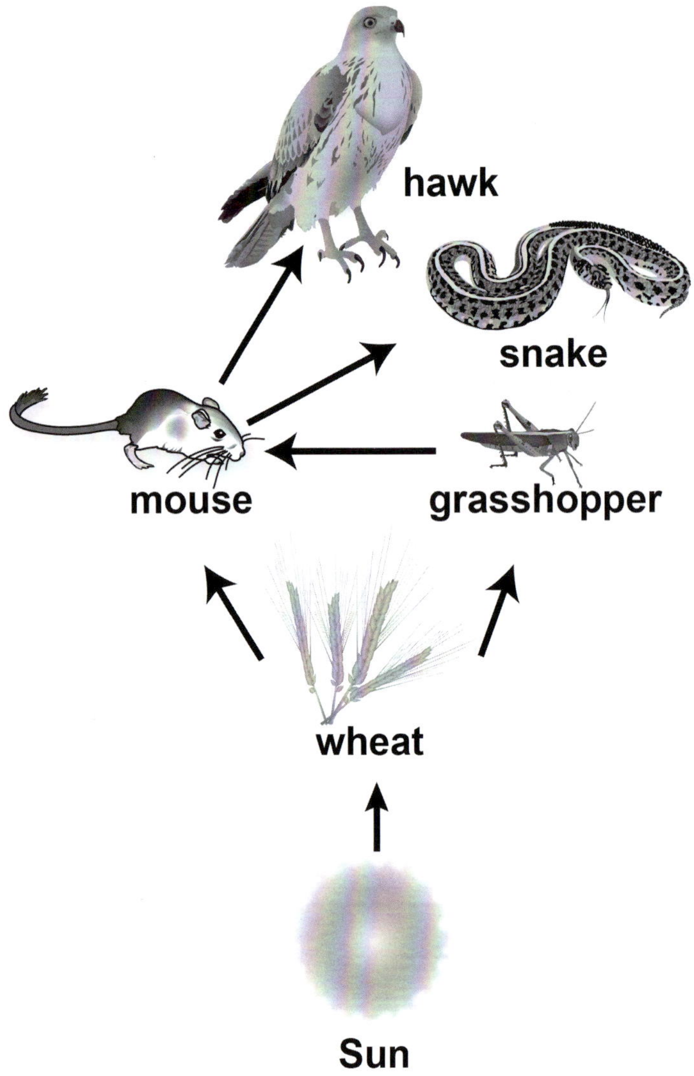

- What is the producer?
- What are the consumers?
- What most likely would happen to the mouse population if the number of snakes declined? Why?

Complex Food Webs

Some food webs have many food chains that overlap.

Observe the complex food web below.

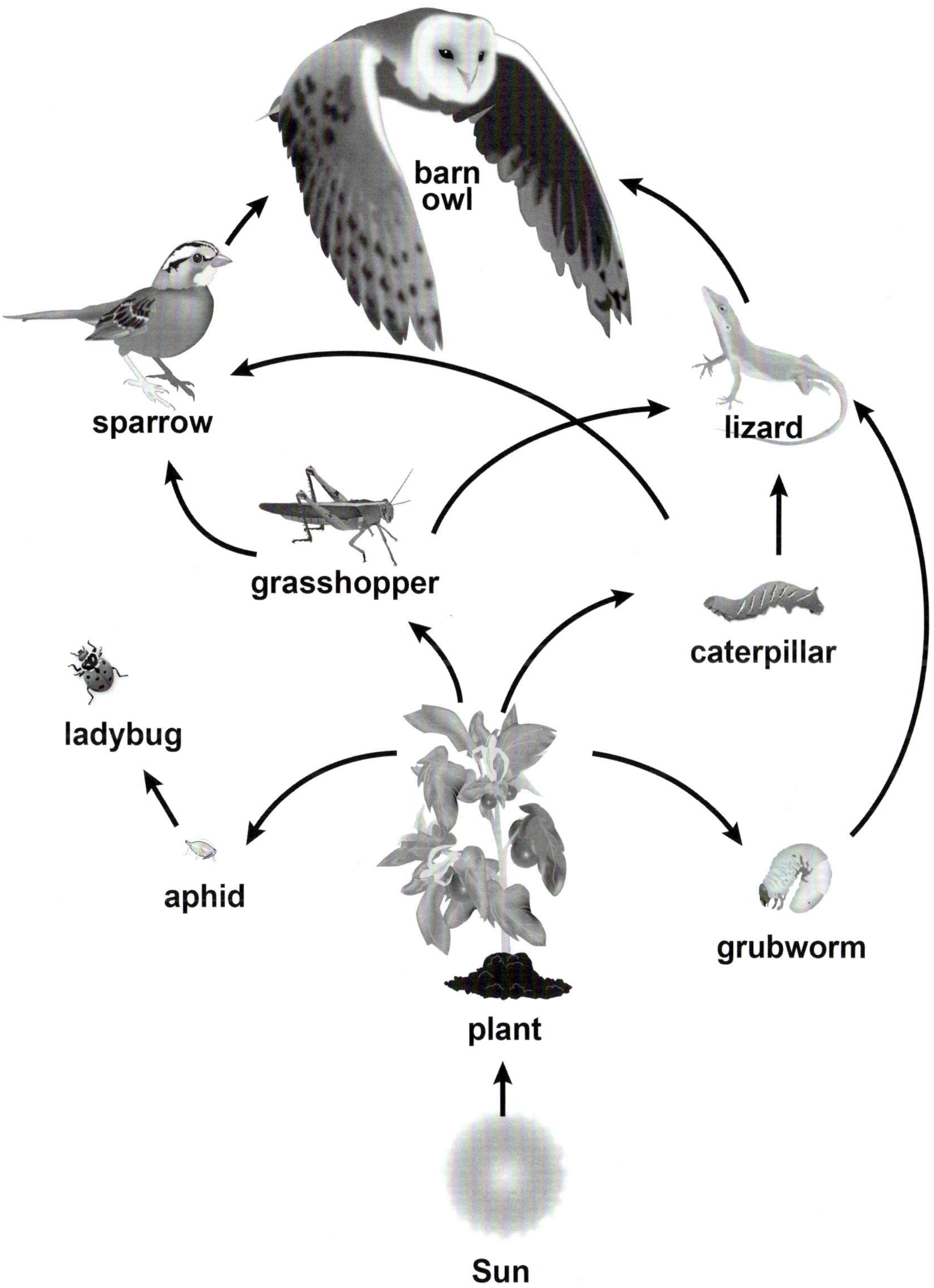

- What is the producer in the food web?
- How do you know?

© REGION 4 EDUCATION SERVICE CENTER. ALL RIGHTS RESERVED.

- What does a lizard eat?
- What does a sparrow eat?
- In what type of ecosystem would this food web be located?

In your science notebook, write two questions that can be answered by observing the complex food web. Include your answers.

Let's Evaluate!

Use your knowledge of food chains and food webs to complete the assessment.

Let's Engage!

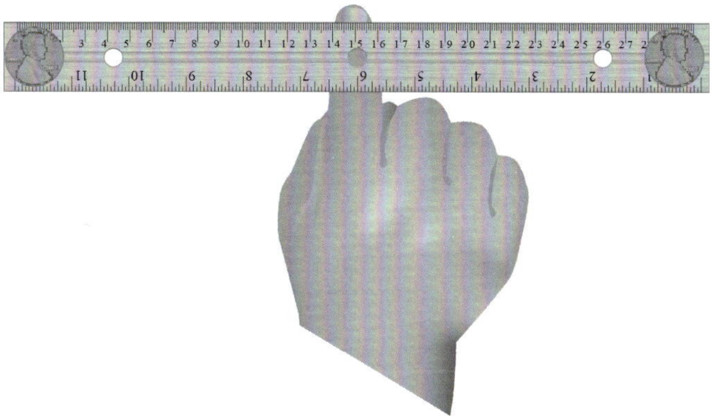

Work as a team to re-create the system in the picture.

- How do you keep the system balanced?
- What happens to the system when a part is missing, such as a penny?

Let's Explore!

Follow your teacher's instructions.

Collect and record accurate data.

Think about—

- What basic need of living things are you exploring?

Let's Explain!

Animal Respiration

Animals need energy to move, grow, and reproduce. They do not make their own food, as plants do. Animals eat plants and other animals to get the energy they need. The process of using food and oxygen to release energy is called respiration.

© REGION 4 EDUCATION SERVICE CENTER. ALL RIGHTS RESERVED.

Breathing is a part of the process of animal respiration. Animals breathe in, or inhale, oxygen from the air. As they breathe out, or exhale, carbon dioxide is released back into the air.

Think about—

- The jumping jack activity.

You counted the number of breaths you took in 1 minute before beginning to exercise. As you exercised, the number of breaths you took per minute increased.

Oxygen was necessary for more energy to be released. Your breathing became stronger and deeper. It was necessary to inhale more oxygen for you to complete the exercise.

Making the Connection

Remember, air is a mixture of gases. Oxygen is one of those gases.

Where does the oxygen in the air come from? Think about the process of photosynthesis. Plants release oxygen during photosynthesis. Animals then inhale the oxygen during respiration. When animals exhale, carbon dioxide is released back into the atmosphere.

What organisms need carbon dioxide? The leaves of plants absorb carbon dioxide to be used for photosynthesis. During photosynthesis, plants make their food and release oxygen into the air.

What happens to the oxygen? Animals need oxygen in order to live and grow. Animals breathe the oxygen released into the air by photosynthesis.

This exchange of gases forms an important cycle for life on Earth.

Observe the cycle.

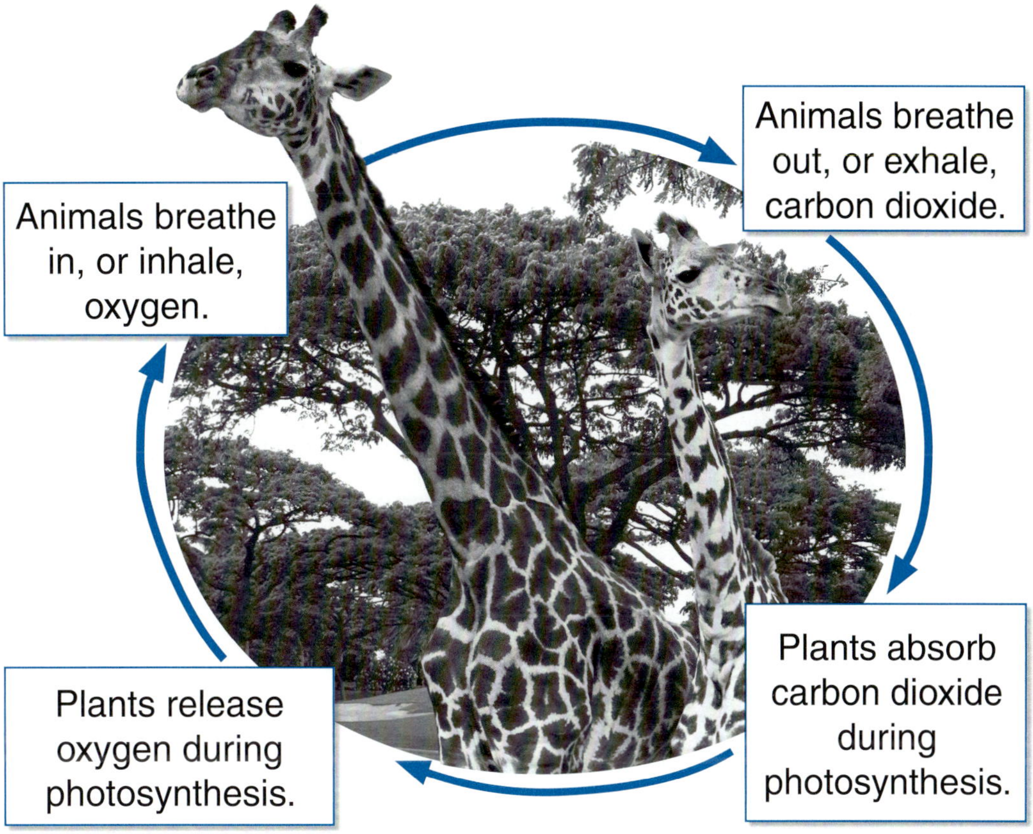

Part 1

1. What types of animals live in the area around your school?
2. What types of plants grow in the area around your school?
3. Draw a diagram of the carbon dioxide-oxygen cycle using plants and animals that are found near your school.

Part 2

Discuss the following questions with your group. Record your answers in your science notebook.

1. What would happen to the carbon dioxide-oxygen cycle if animals ceased to exist?
2. What would happen to the carbon dioxide-oxygen cycle if plants ceased to exist?

© REGION 4 EDUCATION SERVICE CENTER. ALL RIGHTS RESERVED.

Let's Elaborate!

The Effects of Change in an Ecosystem

An ecosystem includes living and nonliving parts. These parts work together to meet the basic needs of the organisms in the ecosystem. When this happens, the ecosystem is balanced.

Sometimes an ecosystem becomes unbalanced. The relationships between organisms are interrupted. The ecosystem can no longer meet the basic needs of its living organisms.

Think back to the ruler system. What caused the system to become unbalanced? The system became unbalanced when a part of the system was removed.

Who's Left?

1. Select a card from the container.
2. Follow your teacher's instructions.
   - In what ecosystem did the animals live?
   - What were the parts of the ecosystem?
   - What happened when a part of the system was removed?

Causes of Environmental Change: Natural Events

Ecosystems are part of the environment. When changes occur in the environment, ecosystems can be affected. Natural events can cause changes in the environment.

Drought occurs when the amount of rainfall over an extended period of time is less than expected. Farmers may not have enough water for their crops. Cities may limit the amount of water used for watering lawns and gardens.

Water is a basic need of living things. Lack of water in an ecosystem causes the ecosystem to become unbalanced. Without rainfall, producers do not survive and grow. Food chains are disrupted. Animals move to other locations to find food and water.

Flooding occurs when there is more water than lakes, ponds, and rivers can hold. A flood may be a result of large amounts of precipitation that overflows into areas that usually are not covered by water. This destroys habitats, causing animals to relocate.

Causes of Environmental Change: Humans

Human interference can cause changes in the environment.

Fire can bring about widespread destruction to our environment, affecting ecosystems. Carelessness of humans is one way that fires may start. The simple act of throwing a lit cigarette out a car window can ignite and cause a fire. A fire also can be the result of campers leaving hot ashes at a campsite.

Fire destroys thousands of acres of land every year. The ecosystems in the path of a fire are damaged, if not completely destroyed.

The growth of towns and cities also can cause changes in the environment. More homes and businesses are needed as the population of a town or city increases. Why does this happen? Humans are living organisms. A basic need of humans is shelter. As the human population of an area increases, homes, businesses, and roads are built. Many times, new construction is located in unspoiled, natural areas.

Think about—

- Most animals leave the area in order to survive.

© REGION 4 EDUCATION SERVICE CENTER. ALL RIGHTS RESERVED.

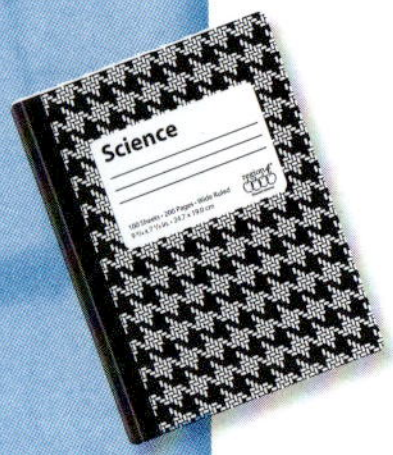

1. Read the statement below.
    - A highway is being built between two towns. The highway cuts through a woodland area.
2. Predict how this would affect the environment. Be specific.
3. Record your predictions.
4. Be ready to share your ideas with the class.
5. Think about—
    - What could humans do to lessen their impact on the environment?

Let's Evaluate!

Use your knowledge of balanced environments and causes of change in ecosystems to complete the assessment.

Unit 7

Organisms and Environments, Part 2

UNIT 7: Organisms and Environments, Part 2

Introduction

Do you have brown eyes? Can you curl your tongue? These are inherited traits. An inherited trait is a characteristic that is passed from parents to their offspring.

Does a newborn puppy know how to play fetch? No, a puppy is taught to run after a thrown stick and bring it back. This is called a learned behavior.

Do you look exactly the same now as you did in kindergarten? Of course not. You have changed. You have grown. All living organisms, including plants and animals, grow and go through changes during their lives.

By the end of this unit, you will—

Know these concepts:

- Adaptations are inherited traits that help an organism survive in its environment.
- Inherited traits in plants and animals are passed from parent to offspring.
- Learned behaviors must be taught and practiced.
- A life cycle includes the stages of growth and development of a living organism.
- Some organisms change through a process called metamorphosis.

Be able to do these things:

- Make scientific observations.
- Collect and record scientific data.
- Make models to represent the natural world.
- Identify structures that help animals survive in their environment.
- Identify traits that are inherited.

- Give examples of learned behaviors.
- Identify and describe the differences between complete and incomplete metamorphosis of insects.

Use these tools:

funnel

graduated cylinder

calculator

hand lens

camera

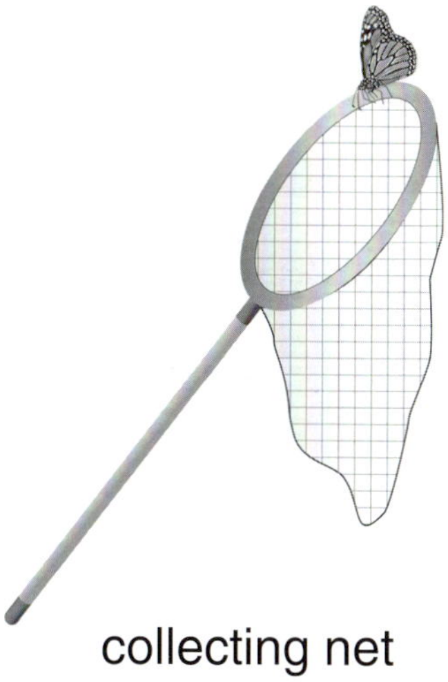

collecting net

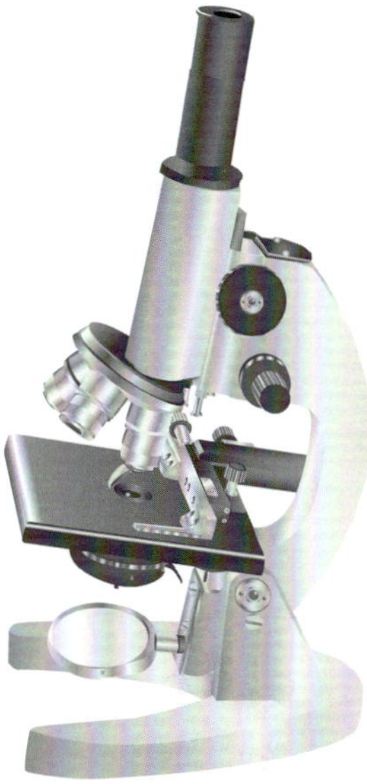

microscope

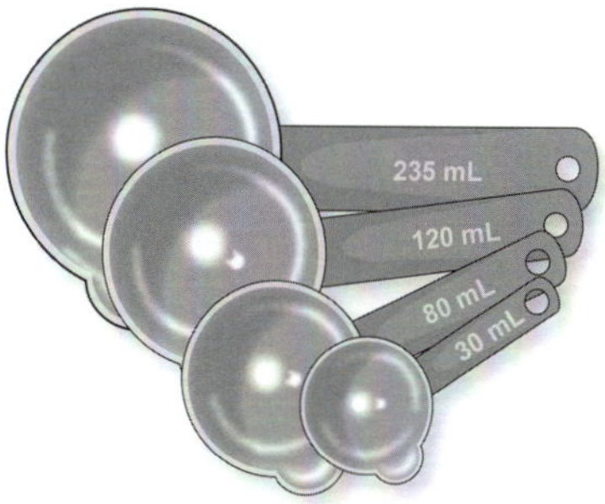

metric measuring cups

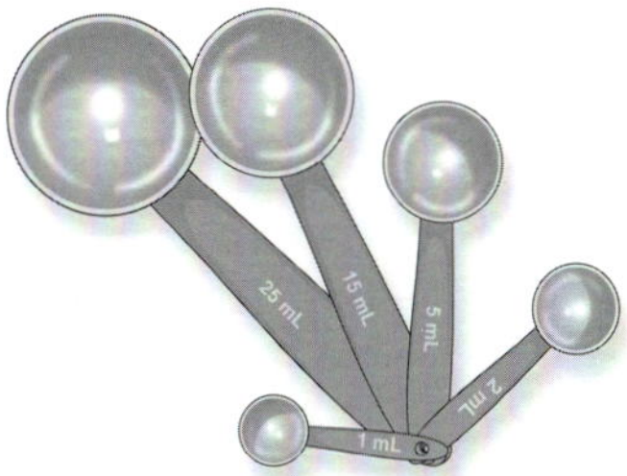

metric measuring spoons

© REGION 4 EDUCATION SERVICE CENTER. ALL RIGHTS RESERVED.

UNIT 7: Organisms and Environments, Part 2

Lesson 1: Structures and Functions of Organisms

Let's Engage!

Try this:

1. Place a pencil on your desk.
2. Make a fist with your hand.
3. Try to pick up the pencil with your fist.
    - Are you able to write your name with the pencil?
4. Open your fist.
5. Try to pick up the pencil.
    - Are you able to write your name with the pencil?

Let's Explore!

1. Copy the table into your science notebook.

Body Structure	Amount of Water (mL)			
	Trial 1	Trial 2	Trial 3	Average
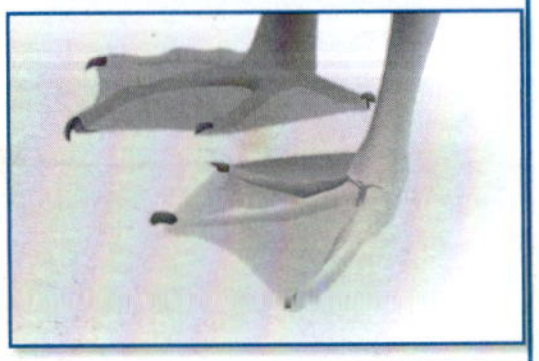				

2. Follow classroom safety rules.
3. Read the lab procedure before beginning the investigation.

4. Follow each step carefully.

5. Make, share, and record observations.

Let's Explain!

Physical Adaptations

Living organisms have body parts or structures that help them live and survive in their environments. These body structures are called physical adaptations. Each body structure has a job or function that helps a living organism survive.

Observe the pictures below.

hummingbird

duck

hawk

Discuss the following questions in your group:

- How are the body structures of the birds similar?
- How are the body structures of the birds different?
- How do the body structures, or physical adaptations, of each bird improve its ability to live and survive in its environment?

© REGION 4 EDUCATION SERVICE CENTER. ALL RIGHTS RESERVED.

UNIT 7: Organisms and Environments, Part 2

Lesson 1: Structures and Functions of Organisms

Learning More about Physical Adaptations

Physical adaptations can look very different from one animal to another.

Ducks have large wings, powerful chest muscles, and hollow bones that allow them to travel long distances. Their feet are perfect for swimming and walking in mud. Ducks also have waterproof feathers. They have a wide, flat beak to gather and hold food. These physical adaptations help a duck survive and live in its environment.

Hedgehogs are small animals that live in forest ecosystems. They are covered with sharp spikes called quills. If a hedgehog is threatened by an enemy or predator, it rolls itself into a tight ball. When a predator tries to bite a hedgehog, the predator gets a mouthful of sharp quills. The sharp quills are a physical adaptation.

Adaptations help living organisms survive and grow in their environments.

Use the example below to describe ways humans use their body parts to live and grow in their environments.

I use my *hands* to *record observations in my science notebook*.

Let's Elaborate!

Memory Game

Object of the Game

Choose matching cards, one card with an animal's picture and another card with its matching physical adaptation.

Rules of the Game

1. Read each card to become familiar with the information on the cards.
2. Shuffle the set of cards.
3. Lay the cards face down on a table in the arrangement shown below.

4. Select who will go first by rolling the dice. The player who rolls the highest number goes first.
5. Turn a card over, leaving it face up in its position. Turn a second card over, leaving it face up in its position. If the two cards form a pair, meaning one animal and its matching adaptation, the player picks up the cards. The player then takes another turn. If the cards do not match, both cards are turned face down. The next player then takes his or her turn.

© REGION 4 EDUCATION SERVICE CENTER. ALL RIGHTS RESERVED.

Winner

The player who finds the greatest number of matching pairs is the winner.

Let's Evaluate!

Use your knowledge of structures and functions of organisms to complete the assessment.

Let's Engage!

How many students in your group—

- Have brown eyes?
- Are female?
- Can roll their tongues?

Be ready to share your observations.

Let's Explore!

What Will It Look Like?

1. Copy the table below into your science notebook.

Flower Part	Physical Trait
petals	
center	
leaves	
stem	

© REGION 4 EDUCATION SERVICE CENTER. ALL RIGHTS RESERVED.

2. Choose one flower part from each paper bag.

3. Record in the table the physical traits of the flower parts.

4. Use the physical traits in the table to create an offspring of the parent plants.

5. Use each flower part as a template to draw a picture of the offspring.

6. Think about—

   - In what ways does your offspring look like the parent plants?
   - Do any of the offspring in your group look exactly like a parent plant?
   - Are any of the offspring in your group exactly the same?

Let's Explain!

Inherited Physical Traits

An inherited trait is a characteristic that is passed from parents to offspring. Parents cannot choose which traits are passed to their offspring.

Inherited traits can be observed in both plants and animals.

A plant's offspring are its seeds. Seeds contain physical traits from each parent plant. A seed grows into a new plant that is similar to the parent plants.

For example—

An orange is picked from an orange tree. A seed is removed from the orange. It is planted and a new tree grows. The new tree has the same shape of leaves and type of bark as the parent tree.

Inherited physical traits of plants include:

- stem length
- bark type
- petal color
- leaf shape

The offspring of animals may be born alive, such as a dog giving birth to puppies. Some animals hatch from eggs, such as baby chicks hatching from chicken eggs. Either way, the offspring look similar to their parents.

For example—

The offspring of a grizzly bear is called a bear cub. The bear cub looks like its parents. Both the parents and offspring have the same number of arms, legs, and toes. They also have the same ear shape and fur color.

Inherited physical traits of animals include:

- hair color
- eye color
- nose shape
- number of toes and fingers

Inherited Behaviors

Offspring may also inherit how they act or behave. These traits may be referred to as inherited behaviors, or instincts.

Think about—

- How would isopods in a terrarium respond if they were touched? Did someone teach the isopod to roll into a ball when threatened?

© REGION 4 EDUCATION SERVICE CENTER. ALL RIGHTS RESERVED.

Lesson 2: Inherited Traits and Learned Behaviors

- What would earthworms do if they were placed on top of the soil in a terrarium? Did someone teach the earthworm to burrow into the soil?

No one taught the isopod to roll into a ball when threatened. No one taught the earthworm to burrow into the soil. These actions are inherited behaviors. The animals instinctively know how to act.

For example—

A baby bird pecks its way out of its shell at just the right time to be able to survive. The pecking of the baby bird is an inherited behavior or instinct.

There are many types of inherited behaviors.

Ducks travel to warmer areas for the winter months and return in spring. This journey is called migration. Ducks do not learn to migrate. They are born knowing when and how to migrate. Migration is an inherited behavior.

Hedgehogs are nocturnal. They like to hunt for insects and worms at night, not during the day. A hedgehog uses its claws and pointed nose to search for food under dry leaves. If food gets scarce in the winter and the temperatures drop, hedgehogs hibernate by going into a deep sleep. When the temperatures warm up and the food supply increases in the spring, hedgehogs come out of hibernation. Hibernation helps hedgehogs survive harsh conditions. Hibernation is an inherited behavior.

The coral snake is poisonous. It has red, yellow, and black bands of color that make it very easy to see. Most animals leave this brightly colored snake alone to avoid its poisonous bite. The same bright colors can be found on the scarlet king snake. This snake is harmless, but predators avoid it as well. One animal copying or imitating another animal in appearance or behavior is called mimicry. Mimicry is an inherited trait that helps animals live and survive.

Some animals have adaptations that help them blend into their surroundings. This is called camouflage. Walking sticks are insects that look just like a limb of a tree.

Photograph by Shelly D. Smith

When the walking stick moves slowly along the limb of a tree, predators cannot see it. A walking stick has the same color, texture, shape, and pattern as a tree limb.

© REGION 4 EDUCATION SERVICE CENTER. ALL RIGHTS RESERVED.

Learned Behaviors

Not all behaviors are inherited. Some behaviors are learned.

Think about—

- Could you ride a bike when you were born? Could you write your name when you were born? Of course not. Riding a bike and writing your name are learned behaviors.
- A boy walks through the forest and picks up a poison ivy vine. Later he gets a painful, burning rash. Will he pick up poison ivy again without wearing gloves? No, he has learned that poison ivy gives him a rash that hurts.
- Does a newborn puppy understand the command to sit? No, someone teaches a puppy to sit when it hears a command. This is a learned behavior.

Let's Elaborate!

Inherited or Learned?

Work as a team to classify the cards as inherited or learned.

Let's Evaluate!

Use your knowledge of inherited traits and learned behaviors to complete the assessment.

Lesson 3: Complete and Incomplete Metamorphosis

Let's Engage!

Put the photographs in order from youngest to oldest.

- What observations will you make to help you put the photographs in order?

Let's Explore!

Observing a Life Cycle

larva

1. Read the procedure on the activity card.
2. Remember to handle live organisms gently and with care.
3. Complete the activity as a group.
4. Clean your work area.
5. Copy the table on the next page into your science notebook.
6. Leave room on the page for extra days.
7. Record your observations for Day 1.

© REGION 4 EDUCATION SERVICE CENTER. ALL RIGHTS RESERVED.

Day	Stage of Life Cycle	Observations
1		
5		
10		
15		
20		

Care of Habitat

- Do not add water to the container. The food mixture supplies enough moisture for the larvae.
- Check the temperature of your classroom daily. Wax moth larvae grow best in temperatures between 22 °C and 26 °C.

Let's Explain!

Life Cycles

You already know that—

- Offspring inherit characteristics from their parents.
- Parents cannot choose what traits they pass to their offspring.
- Living organisms grow and change over time.

The stages of growth of a living organism are called a life cycle. Not all life cycles are the same. Some offspring look like their parents when they are born. Others begin life in one form and over time develop into something completely different.

The change in the physical form of a living organism in its life cycle is called metamorphosis.

Complete Metamorphosis

The life cycle of a wax moth has four distinct stages. At each stage, the insect changes form. This is called complete metamorphosis.

Observe the pictures of the life cycle of a wax moth.

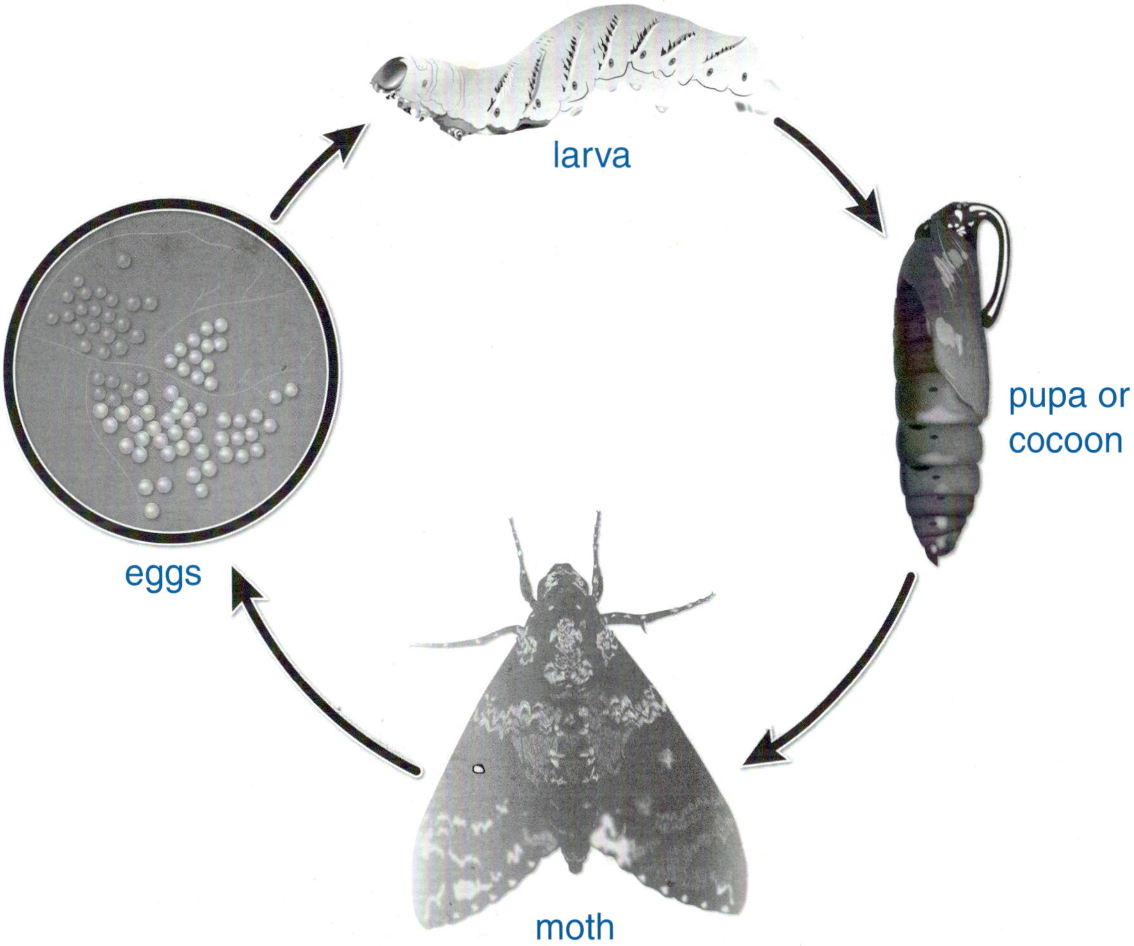

The first stage is a tiny egg. An adult wax moth will lay up to 300 eggs. A hand lens or microscope is needed to observe this stage of the life cycle.

In 5 to 8 days, wax moth eggs hatch. The second stage is called the larva. The larva is white in color and usually grows to about 2 cm in length. Wax moth larvae have fatty bodies. Lizards, hedgehogs, and fish like to eat the larvae.

© REGION 4 EDUCATION SERVICE CENTER. ALL RIGHTS RESERVED.

A wax moth larva will eat for 3–4 weeks and change into a pupa, or cocoon. This is the third stage of its life cycle.

The fourth and final stage of the wax moth life cycle is the adult moth. A wax moth emerges from the cocoon in about 2 weeks. It lives about 3 weeks and lays about 300 eggs. The life cycle continues.

Lesson 3: Complete and Incomplete Metamorphosis

The life cycle of a monarch butterfly is very much like that of a wax moth. The third stage of the butterfly life cycle, however, is a chrysalis, not a cocoon.

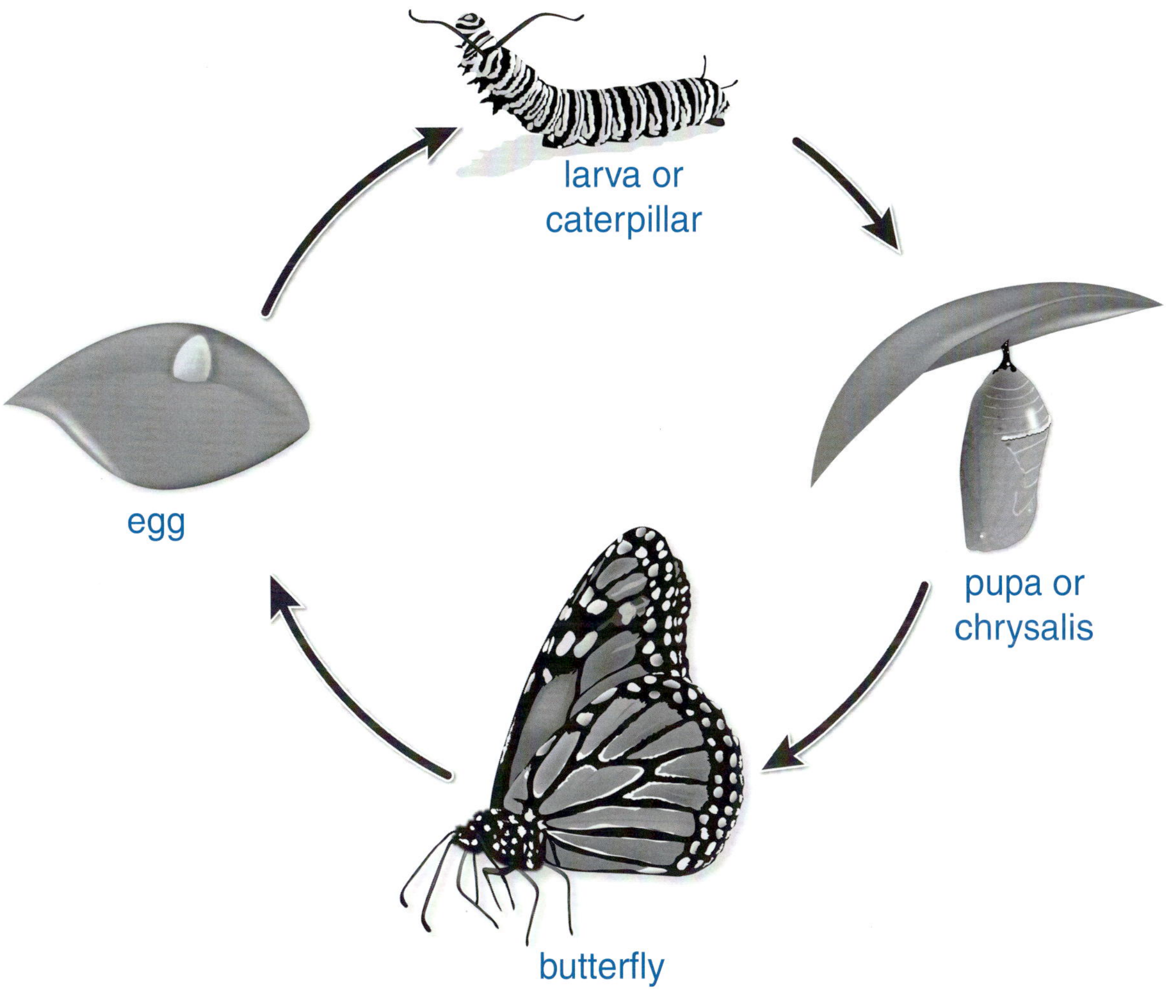

Moths and butterflies go through complete metamorphosis. Other insects that go through complete metamorphosis include beetles, flies, fleas, wasps, bees, and ants.

- Choose an insect that goes through complete metamorphosis.
- Draw and label the four stages of its life cycle.

© REGION 4 EDUCATION SERVICE CENTER. ALL RIGHTS RESERVED.

Let's Elaborate!

Incomplete Metamorphosis

Observe the life cycle of a grasshopper.

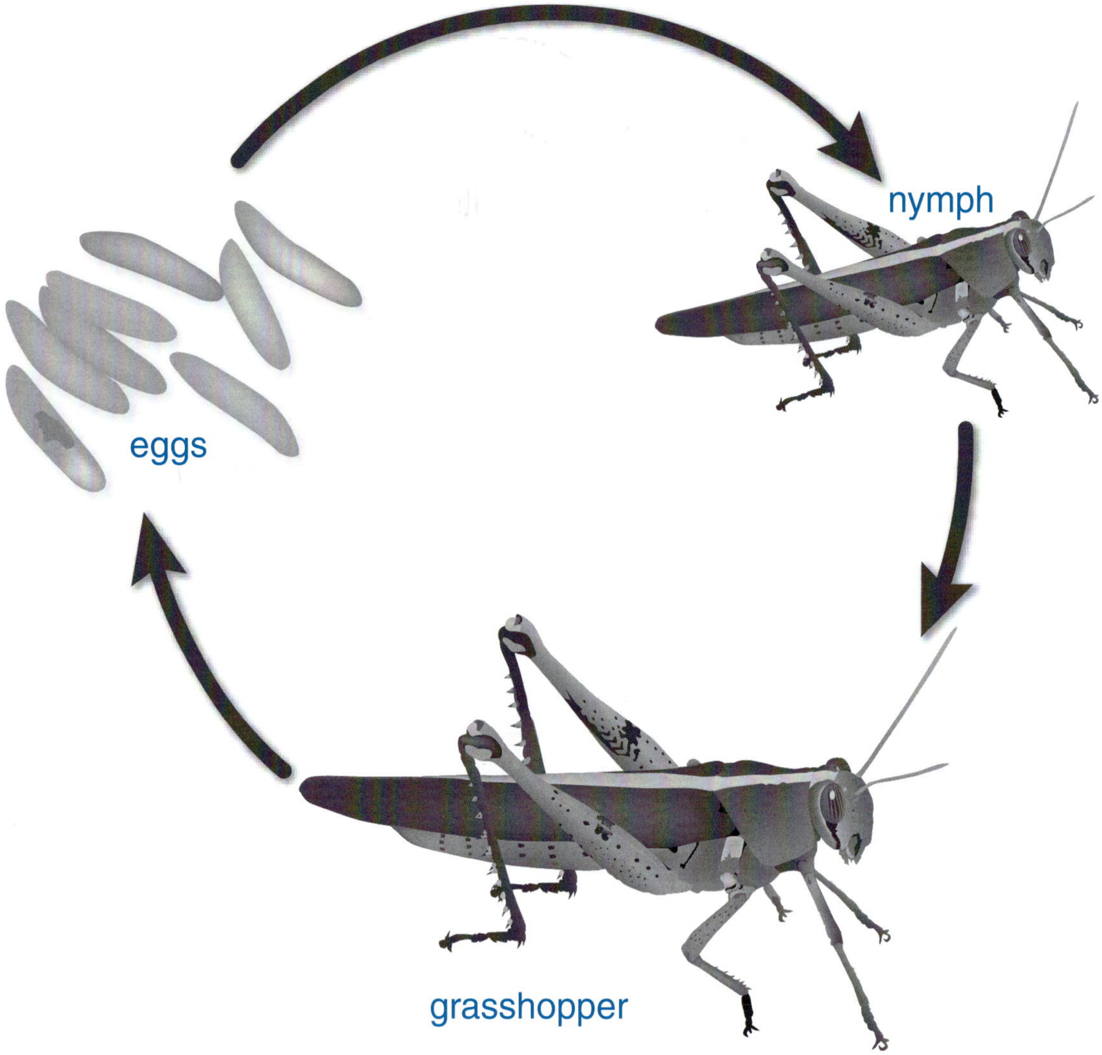

Compare the life cycle of a grasshopper to the life cycle of a butterfly.

- How are the cycles similar?
- How are the cycles different?

Lesson 3: Complete and Incomplete Metamorphosis

A grasshopper grows and develops through three stages in its life cycle, but its form does not appear to change. The nymph, however, is a wingless version of the adult insect. As the nymph ages, it grows wings. This type of growth is called incomplete metamorphosis.

Other insects that go through incomplete metamorphosis include crickets, cockroaches, and head lice.

- Choose an insect that goes through incomplete metamorphosis.
- Draw and label the three stages of its life cycle.

Let's Evaluate!

Use your knowledge of complete and incomplete metamorphosis to complete the assessment.

© REGION 4 EDUCATION SERVICE CENTER. ALL RIGHTS RESERVED.

Appendix

Science Notebook

Guidelines

A science notebook is much more than just a regular notebook. It is a place where you record your experimental data, make notes from your research and discussions, and write reflections and summaries about what you observe and learn. Your science notebook is a reflection of what you know and are able to do in this science course. It provides a visible, permanent record of your thinking and learning.

Drawings often explain more than words. Throughout your science notebook, you should have many diagrams, sketches, and drawings. Graphic organizers such as Venn diagrams and concept maps help organize your thoughts or material that is being learned.

You are expected to follow these guidelines and any others presented by your teacher.

1. All writing is neat and legible.
2. Pages are not lost or torn out.
3. Mistakes are erased or drawn through with one line.
4. All science notebook entries begin with the appropriate date.
5. Creativity, curiosity, and thinking are evident.
6. Artwork is relevant to the topic and tasteful in nature.

Element	Points Possible	Earned Assessment	
		Self	Teacher
All science notebook entries are dated.			
Diagrams, drawings, and sketches are used to explain concepts and methods.			
Graphic organizers are used to organize thinking and material being learned.			
Problems and concerns are identified and possible solutions are suggested.			
Curiosity and creativity are evident.			
Concepts are supported with accurate details.			
Appropriate science vocabulary is applied.			
Entries are neat and presentable.			
TOTAL			

© REGION 4 EDUCATION SERVICE CENTER. ALL RIGHTS RESERVED.

Drawings and Diagrams

Guidelines

Drawing pictures and diagrams is a powerful way to demonstrate what you know and want to remember. You do not have to be a talented artist to make good scientific drawings. By following some simple steps, you can make diagrams that help you and others know what you have learned.

Steps:

1. Use straight lines to label the parts of the diagram.
2. Never allow label lines to cross each other.
3. Write beside the label line rather than on the label line.
4. Include enough details to allow the diagram to express the concept.
5. Always check your spelling and accuracy of labels.
6. Use color and shading as appropriate.
7. Write a brief caption beneath or beside the drawing to identify the diagram and provide additional information to the reader.

Example of a Poorly Drawn Diagram of Soil Layers in a Container

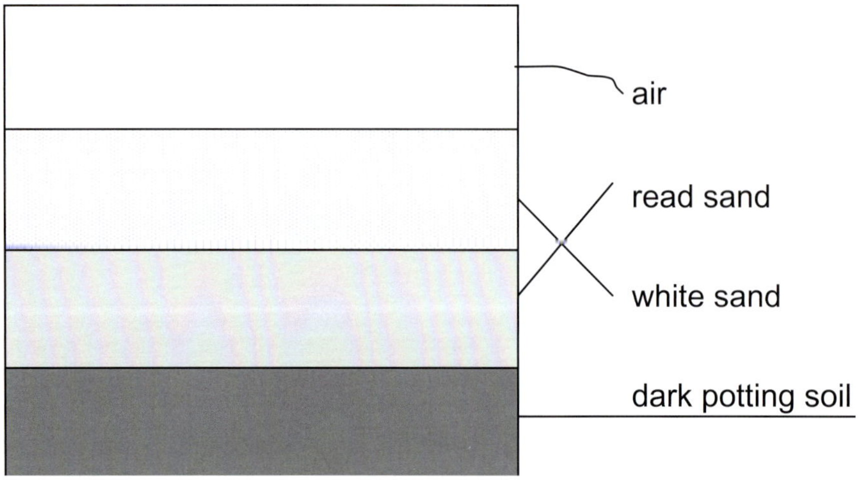

Identify at least five errors made in the diagram.
How would you correct the errors?

Element	Points Possible	Earned Assessment	
		Self	Teacher
The diagram accurately represents the concepts.			
A statement or caption explains what the drawing or diagram is meant to represent.			
Parts are labeled and appropriately named.			
Colors, shading, or other features are used to enhance the pictorial information.			
The drawing uses the space of the paper well.			
If necessary, a key is included.			
All important information is included in the drawing or diagram.			
The drawing or diagram is neat and presentable.			
TOTAL			

© REGION 4 EDUCATION SERVICE CENTER. ALL RIGHTS RESERVED.

Credits

All *Gateways to Science STAAR Edition* series photographs that are listed on these pages are used with permission of the copyright holder. All *Gateways to Science STAAR Edition* series photographs not specifically acknowledged on these pages were produced by Region 4 Education Service Center.

All graphics and diagrams in the *Gateways to Science STAAR Edition* series were developed by Region 4 Education Service Center.

SE indicates Student Edition.
TE indicates Teacher Edition.

Unit 1
Page 7 SE, Page 9 TE (tiki torch): Hanna Roach

Unit 2
Page 19 SE, Page 24 TE (cheese): Shelly D. Smith
Page 23 SE, Page 29 TE (boy being measured for height): Diana Elizondo

Unit 3
Page 85 SE, Page 138 TE (lighthouse): Hanna Roach
Page 85 SE, Page 138 TE (sky, rainbow photo manipulation): Shelly D. Smith

Unit 4
Page 90 SE, Page 145 TE (water cycle): Shelly D. Smith
Page 93 SE, Page 150 TE (sunset over ocean): Shelly D. Smith

Unit 5
Page 114 SE, Page 181 TE (Antelope Island, Utah): Shelly D. Smith
Page 116 SE, Page 185 TE (glacial valley—Sun Point Overlook, Glacier National Park): David Restivo, National Park Service
Page 116 SE, Page 185 TE (glacial erosion—Thunderbird Mountain, Glacier National Park): National Park Service
Page 117 SE, Page 186 TE (Queule, Chile, 2 photos): NGDC, NOAA, US Department of Commerce
Page 118 SE, Page 188 TE (beach, Hawaii): Shelly D. Smith
Page 119 SE, Page 189 TE (Great Sand Dunes National Park and Preserve, Colorado): National Park Service
Page 121 SE, Page 193 TE (Grand Canyon, Arizona): Jennifer Wilkening
Page 124 SE, Page 200 TE (sandstone, limestone, 2 photos): Shelly D. Smith
Page 131 SE, Page 210 TE (Lake Travis, Austin, Texas): Karen Thompson
Page 133 SE, Page 215 TE (fossil, petrified wood, 2 photos): Shelly D. Smith
Page 135 SE, Page 218 TE (fish fossil, top): Shelly D. Smith
Page 136 SE, Page 219 TE (dinosaur skull, dog, dinosaur leg bone, 3 photos): Shelly D. Smith

Unit 6
Page 259 TE (squirrel): Anne Douglas

Unit 7
Page 177 SE, Page 285 TE (walking stick): Shelly D. Smith